You Have to be Lucky

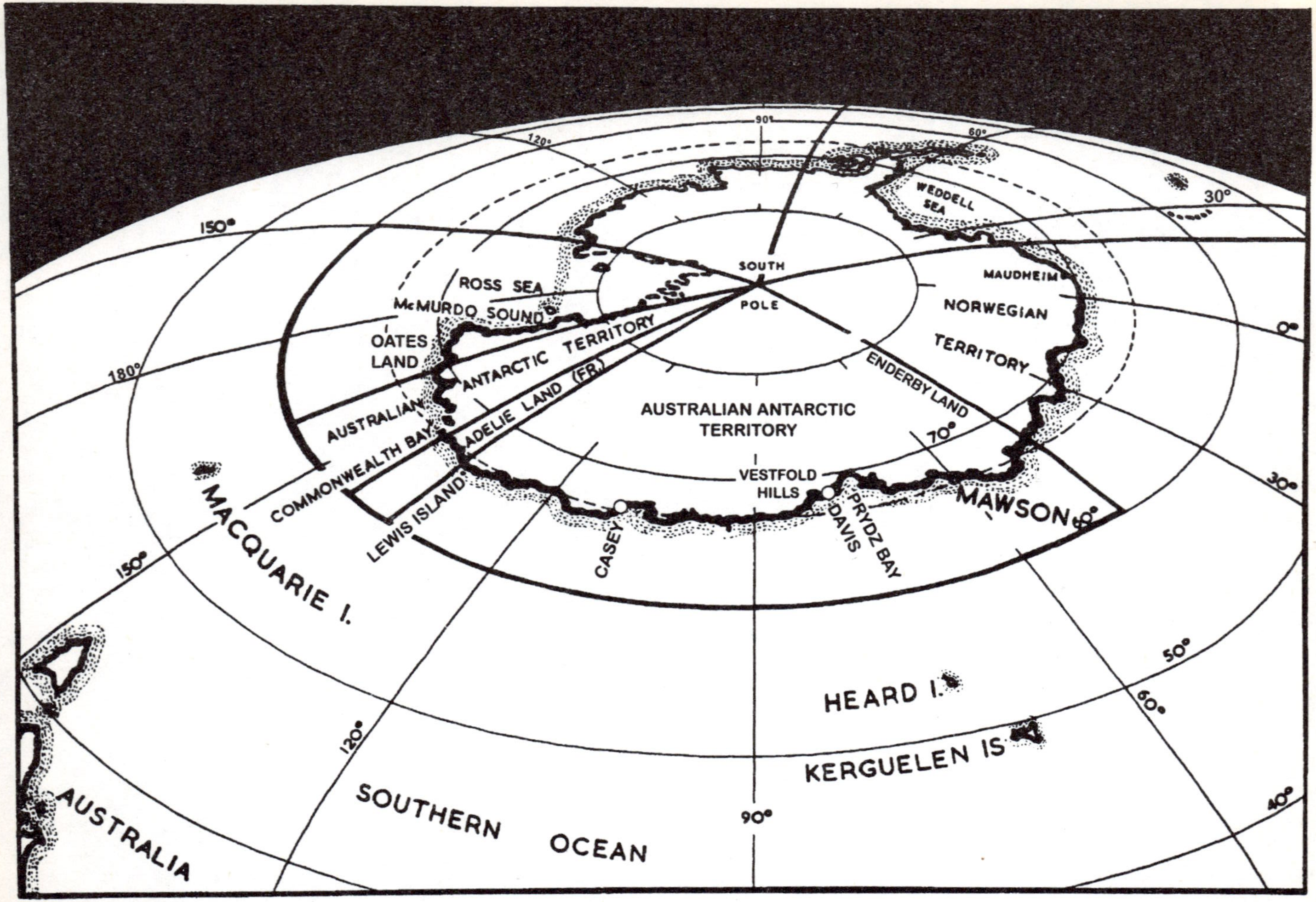
SOUTH POLE
ROSS SEA
McMURDO SOUND
OATES LAND
AUSTRALIAN ANTARCTIC TERRITORY
ADELIE LAND (FR.)
COMMONWEALTH BAY
LEWIS ISLAND
AUSTRALIAN ANTARCTIC TERRITORY
VESTFOLD HILLS
CASEY
DAVIS
PRYDZ BAY
MAWSON
ENDERBY LAND
NORWEGIAN TERRITORY
MAUDHEIM
WEDDELL SEA
MACQUARIE I.
HEARD I.
KERGUELEN IS
SOUTHERN OCEAN
AUSTRALIA
90°
120°
150°
180°
60°
30°
0°
70°
50°
40°

YOU HAVE TO BE LUCKY

ANTARCTIC AND OTHER ADVENTURES

PHILLIP LAW

Kangaroo Press

Other books by Phillip Law

ANARE (Oxford University Press, 1957) with J.Béchervaise

Antractic Odyssey (Heinemann Australia, 1983)

First published in 1995 by Kangaroo Press Pty Ltd
3 Whitehall Road Kenthurst NSW 2156 Australia
P.O. Box 6125 Dural Delivery Centre NSW 2158
Printed by Australian Print Group, Maryborough, 3465

ISBN 0 86417 743 7

Contents

Acknowledgments

This book was written during the five-year period that elapsed between the stroke suffered by my wife Nel and her ultimate death. Looking after her in our home made such demands upon my time that I had difficulty in attending to my writing and a number of other commitments.

Upon the initiative of Ian McRitchie, the Phillip Law Trust Fund was established under the auspices of the Museum of Victoria, with major contributions from several trusts and donations from a number of individuals. To all of these I am greatly indebted and particularly to the John T. Reid Family Trust, the Sidney Myer Foundation, the Perpetual Trustees Fund, the R. E. Ross Trust, the Dank Trust, the Balwyn Rotarians, Robert and Betty Lewis, Geoffrey Law, Dame Elizabeth Murdoch, Rob, Margaret and Peter Vial and others.

The Trust Fund provided me with domestic and secretarial assistance that enabled me to write this book and another (recently published), to continue to work towards the establishment of an Antarctic collection within the Museum of Victoria and to embark upon a project aimed at purchasing the historic Antarctic ship *Kista Dan* and installing her in the Port of Melbourne as a museum ship.

I thank journalist Sally White for her valuable editorial advice, which lead to a considerable revision of my first draft of this book. I also acknowledge the valuable work of my part-time secretary, Julie Di Pietro, during this period, including the word-processing of this manuscript.

Beyond Kerguelen

Henry Lawson *(Selected stanzas)*

Down in the South, by waste without sail on it,
    Far from the zone of the blossom and tree—
Leith, with winter and whirlwind and wail on it,
    Ghost of a land by the ghost of a sea.
Weird is the mist from the summit to base of it;
    Sun of its heaven is wizened and grey;
Phantom of light is the light on the face of it—
    Never is night on it, never is day!
Here is the shore without flower or bird on it;
    Here is no litany sweet of the springs—
Only the haughty, harsh thunder is heard on it,
    Only the storm with a roar in its wings!

Shadow of moon is the moon in the sky of it—
    Wan as the face of a wizard and far!
Never there shines from the firmament high of it
    Grace as the planet or glory of star.
All the year round, in the place of white days on it—
    All the year round where there never is night—
Lies a great sinister, bitter, blind haze on it;
    Growth that is neither of darkness nor light.
Wild is the cry of the sea in the caves by it—
    Sea that is smitten by spears of the snow;
Desolate songs are the songs of the waves by it—
    Down in the south, where the ships never go.

Storm from the Pole is the singer that sings to it
    Hymns of the land at the planet's grey verge.
Thunder discloses dark, wonderful things to it—
    Thunder and rain, and the dolorous surge.
Hills with no hope of a wing or a leaf on them,
    Scarred with the chronicles written by flame,
Stare through the gloom of inscrutable grief on them,
    Down on the horns of the gulfs without name.
Cliffs with the records of fierce flying fires on them.
    Loom over perilous pits of eclipse;
Alps with anathema stamped in the spires on them—
    Out by the wave with a curse on its lips.

INTRODUCTION

LUCK

The concept of luck is universal. We talk of lucky and unlucky people, of a 'run of luck' by a gambler, of 'bad luck' striking down a person in some accident. The Oxford dictionary defines luck as the 'supposed tendency of chance to bring a sequence of (un)favourable events'; and 'fortuitous events affecting one's interests'. So there can be a simple 'lucky event', or a sequence of events producing a 'run of luck'. The superstitious idea of a person being endowed with good luck or plagued by bad luck is generally scorned by scientists, who use statistics to explain such events. There is always *some* probability of the occurrence of a sequence of events of a certain kind, no matter how improbable the occurrence may seem (like, for instance, tossing 'heads' with a coin 20 times in succession).

Nevertheless, most of us *do* believe in lucky (or unlucky) people. The probability argument impresses one in relation to a single lucky event; but a continuing succession of lucky events over a long period, while still of finite statistical probability, reinforces our superstitious beliefs.

Psychologically, too, such beliefs are reinforced. A person who experiences some good luck becomes, as a result, more confident, and any continued good luck increases this confidence. Conversely, bad luck creates despondency, which is increased by any continuation of such luck. In both cases it is more natural that the persons concerned will adopt the concept of luck rather than, unemotionally and objectively, examine the statistical probabilities concerned.

It is often said that one makes one's own luck. It is certainly true that confidence breeds success, whether in sport or business and, in this sense, subsequent good luck may be attributable to the extra confidence generated by an initial lucky break. One needs confidence to take risks and, if the gambles succeed, there is a kind of upward spiralling of confidence that encourages further gambles. One is thus inspired 'to run one's luck'.

Some 'runs of luck' stretch probabilities to unbelievable lengths. Twenty-odd years ago a man won the three major lotteries in Australia within a period of two years—Tattersalls in Victoria, the Golden Casket in Queensland and the NSW State Lottery. People who 'break the bank at Monte Carlo' provide other examples. I knew a man who worked his way through university by betting on anything and everything and almost never losing. He would 'toss you for two bob' whenever he met you, win at the races each Saturday, and even win successions of weekly euchre tournaments at the local church.

At the other extreme, I watched sadly as 'bad luck' struck down a large family one by one, in a sequence of horrible tragedies.

Even while recognising, as a scientist, the cold logic of statistics, I prefer to cling to my emotional superstition that I am a 'lucky' person, mainly because (I think) I gain confidence thereby. In an extrovert sense, too, I find that flaunting this 'lucky' image generates confidence in any people I happen to be leading.

Napoleon was reputed to have avoided appointing 'unlucky' people as generals. I must admit to having a sneaking regard for this principle, but on logical rather than superstitious grounds.

Just as people may 'make their own good luck', so others may 'make their bad luck'. Often a 'failure pattern' attributed to bad luck has arisen as a result of the person's own inadequacies. Viewed in this light, Napoleon's maxim makes sense. I know that I followed the same principle, never appointing to a field party or other dangerous enterprise someone with a history of misadventure. Quite apart from being 'accident prone' because of clumsiness or lack of muscular coordination, some people are 'accident prone' because of thoughtlessness, recklessness, inability to make quick decisions, a tendency to panic when things go wrong, and so on.

I continue to be impressed by the tiny margin between success and disaster. You have to be 'lucky' to survive. The 'unlucky' person is struck down by fate at the first encounter, in the simplest fashion, in an everyday incident that involves no element of real danger. The 'lucky' person goes on and on, with hair's breadth escapes one after the other, to a hearty old age. The life of Sir Hubert Wilkins is a good example. Beginning as a war correspondent-photographer in World War I, he took every imaginable risk in the front lines and survived, being awarded the Military Cross and Bar. He was co-pilot in 1928 of the first aeroplane ever to fly over the North Pole and of the aeroplane that made the first flight over Antarctica. He participated in numerous expeditions in both polar regions and, in 1931, embarked on a crazy venture aiming to navigate a wartime submarine beneath the ice to reach the North Pole. 'Luckily' his crew refused to man the craft just as it was about to dive, knowing that there was no chance of succeeding. Wilkins died of a heart attack in 1958, aged 70 years, when sitting on his bed one morning doing up his shoes.

An interesting aspect of my luck is that it does not extend to financial matters. I will 'run my luck' with weather, with adventure and with almost all aspects of normal living but, once money is involved, you can give me away. I have never won a raffle, let alone a lottery. I have no luck at cards. If I buy shares one day the price drops steeply the next, and if I sell some it is then announced that a bonus issue is to be made.

This, I think, is another example of 'making your own luck' I have never been interested in money, or the making of it. Therefore I have never done enough homework before speculating and, because of my lack of interest, I have not kept watch on the financial scene and have been too unconcerned to jump rapidly at the appropriate time.

My Antarctic colleagues used to refer to the 'Law Luck'. We would sail for days through abominable weather and, when I decided to push south for a landing, the weather would fine up. Although the 'calculated risks' we took were pretty carefully assessed, we came through time after time without misadventure. In everyday life, not only in the Antarctic, I have been surprised to recall on how many occasions I have missed serious injury by a hair's breadth. The following episodes illustrate what I mean.

1

HURRICANE IN PRYDZ BAY

In March 1954 MV Kista Dan *was returning to Australia from Antarctica after establishing Mawson Station and exploring the Vestfold Hills.*

I lay in my bunk, rolling from side to side with the violent motion of the ship, wedged into place as well as I could manage it with my knees pressed against the bulkhead on one side and my back against a rolled-up blanket laid lengthwise against the side board.

Squadron-Leader Doug Leckie called me at 0130 hours on Friday 5 March to tell me that our Auster aircraft, lashed down on the deck, had been wrecked. The wind was now blowing at Force 14 (more than 100 knots) and the ship was hove-to into the wind and rolling and pitching violently in heavy seas.

Doug and I stumbled along the heaving passage and pushed through the door at the end onto the boat deck. It was wild out there. We clung to the handrails as the wind buffeted us and freezing spray lashed our faces. The scream of the wind and the flapping of our parka hoods around our ears made conversation impossible, so I just stood and looked. The aircraft certainly was a mess. Gusts of wind under the starboard wing had caused the main strut, connecting the port float to the fuselage, to collapse and the plane had lurched over on one side, crumpled and twisted, with the port wing tangled up in the lifeboat davits.

There was nothing we could do except to hope that it would remain where it was. With the ship hove-to, there was a reasonable chance that it would. So we retreated inside and climbed the companionway to the bridge to inform the captain of the accident.

Captain Hans Christian Petersen looked grim. He was having great difficulty holding the ship into the wind. Relieved of her cargo, *Kista Dan* was floating high in the water and, as each gust struck her starboard quarter, she would shudder and shy off to port, only to be hauled back into the wind by the helmsman. The wind was now blowing at hurricane force, well over 100 knots. The pitching ship would bury its bows into each giant wave and send a sheet of water and spray slashing over the foredeck and across the windows of the bridge. Spray was freezing on the deck and the rigging and the windscreen-wipers on the bridge windows were clogging with sludge ice.

There was little conversation on the darkened bridge. The helmsman's repetition of the captain's steering orders and an occasional comment in a low voice by the first mate or myself were the only words spoken. The radar shone green under its hood and the first mate seldom left it. The crewman operating the depth sounder watched silently as the recorder swept around, sketching a profile of the sea floor, while out on each wing of the bridge a sailor, huddled up in waterproofs and mittens, peered out into the darkness on the lookout for ice floes and growlers.

The bows were riding so high out of the water that I asked the Chief Engineer whether the forward tanks had been ballasted. He said no; they had tried, but the pipes had frozen up.

A ship returning without cargo from a long voyage can be ballasted by pumping sea water into empty fuel tanks and, at a pinch, into the depleted fresh water tanks. Fuel can also be transferred between various tanks to trim the ship. In *Kista Dan* the fuel in the deep tank forward had been used, as also had much of the fresh water from the forepeak tank. If the pipes had not frozen up, sea water could have been pumped into these tanks, or fuel could have been transferred from the double bottom tanks to the deep forward tank.

About 0200 hours the captain lost control of the ship. No combination of rudder and engine revolutions could counteract the force of the wind on the bows and *Kista Dan* broached to, with her starboard side to the wind. Lying over on her port side, she drifted helplessly, pounded by every breaking wave and held in a permanent list by the hurricane.

Realising that in this position the aircraft would be exposed, I again ventured out onto the boat deck and about 0300 hours I watched dejectedly as another spar collapsed and the plane toppled forward onto its nose on the deck. The propeller was bent back under the fuselage and the starboard wing was crumpled up in a horrible, grinding mess. Shortly afterwards, the wind blew the whole lot over the ship's side and the second mate cut the lashings to let the wreckage drift away. As he did so, petrol from the fuel tank spilled over his hands and he writhed in agony as his fingers froze up. (As petrol does not freeze until its temperature is very low, it can become much colder than water.)

From this point on it was a nightmare. *Kista Dan* lay on her side, held to a permanent list of 30° by the force of the wind and rolling from there, so that the limits of the rolls were vertical or 10° to starboard, then 60° or 70° to port.

We knew the stability of the ship was marginal. In earlier voyages I had been used to very stiff, highly stable polar vessels with rolling periods of five to eight seconds.

Kista Dan during this voyage had demonstrated a rolling period of 14 seconds. Back in December, on the voyage to Macquarie Island, I had written in my diary: 'The ship feels top-heavy; she has a long period of roll and tends to hang over at the end of the roll. It makes for a comfortable ride, but I shall not feel very safe if she hangs over like this in a big storm. (After she returned to Denmark at the end of the voyage the owners added 70 tons to her keel.)

I have never been so terrified. A gust of wind and a huge wave would strike the ship and she would heel over to 70°. There she would hang and shudder as though debating whether to continue on or to return. For what seemed like a minute I would hold my breath, saying 'No, no!', with my heart thumping against my ribs; then she would lurch back to the 30° mark, only to be hurled over again to lie flat on her side. When in my bunk, I was lying on the bulkhead for half the time instead of on the mattress.

There have been a number of occasions in my life when I have been afraid, but never have I experienced, either before or since, terror extending unrelieved over such a long period of time. I don't think anyone on the ship really expected to survive and, over a period of 12 hours, each large roll seemed likely to be the last.

There was nothing for me to do. I felt superfluous on the bridge and was reluctant to intrude with unnecessary comments or questions. Most of my men were in their bunks and, although this was far from comfortable, it was much more comfortable than trying to maintain balance standing up, so I followed their example, only rising periodically to gaze through the porthole or occasionally to show my face for a minute or two on the bridge.

The most horrible moment of the whole night was when I stood in my cabin, in semi- darkness, leaning against the bulkhead, with my face pressed against the porthole, peering out on the port side over the tormented water. All was confusion: the near- darkness, the spray slashing the glass, the blown spume streaking the tossing surface of the sea, the crashings and bangings around the ship, the high-pitched scream of the wind, and the violent movements as the ship reared and plunged like a frightened horse. Suddenly, struck by a fierce gust of wind and a particularly high wave, the ship heeled over until I was lying on my stomach on the bulkhead, practically horizontal, with my nose flattened against the submerged glass of the porthole, gazing down into the green-black depths. There I waited and waited for the recovery that would not come, as successive waves ground the ship further and further down on its side and my terror mounted. Finally, after shaking and shuddering and deliberating, the vessel slowly moved back to the 30°-list position and I stood once again on the deck, with pounding heart and sweat pouring down my face.

In my imagination I could feel the ship go over again, shuddering at the horizontal mark and then turning right over. I could hear myself saying 'This is it!' as I was tossed onto the ceiling along with every loose article in the cabin. I could imagine the roar of water rushing into the ship and rising steadily around me as I floundered to reach the door of my cabin. Beyond that I was not sure. Would I struggle futilely against the flood, or would I resign myself to die slowly as the water rose higher and higher against my chest? Few prospects can be more horrible.

I recall quite clearly my emotion when confronting what then seemed like certain death. It was one of intense anger and resentment. Here I was, the successful expedition leader, returning home with a story of accomplishment and a load of important scientific records and photographs. Was all this to be lost? Was all our work to be wasted? And no-one would ever know exactly what had happened to us!

When broached to, a ship cannot be steered—it simply drifts sideways at the mercy of the wind and the seas. Any attempt to bring the bows around into the wind is defeated by the force of the wind on the hull and superstructure. But the ship is not entirely out of control; she can still proceed forward or astern, driven by the engines, so she can crab around obstacles as she drifts sideways.

And this is what Captain Petersen did over the next 24 hours. We were now well out from the eastern coast in Prydz Bay, which was about

220 kilometres from east to west at that part. The western boundary was defined by the vertical ice cliffs of the Amery Ice Shelf. There were a lot of icebergs and many bergy bits—pieces of floating ice varying in size from that of a room to that of a fair-sized house. If a large wave were to dash even a small bergy bit against the hull of the ship, other than the very strong bow section, it would punch a great hole in its side.

The trouble was to spot the bergy bits. Most of their volume is submerged and, in a rough sea at night, it is very hard to distinguish in the rays of a searchlight between the white crest of a breaking wave and the white top of a bergy bit in that crest. Also, radar does not distinguish between the general clutter of breaking seas close to the ship and any bergy bits floating in them.

There would be minutes of anxiety each time the radar indicated icebergs on the port side. As the ship drifted down on them (a ship drifts faster in the wind than an iceberg) the captain would do his crab act—full speed ahead to drift past the northern end of one berg, then full speed astern to drift past the southern end of a second one, and so on. Then the engineer would warn that, with engines racing on these exercises and with the propeller tips emerging periodically from the water as the ship pitched, the engines were becoming dangerously overheated. The captain would then have to reduce the engine revolutions and the chances of avoiding the icebergs would diminish.

Day dawned dimly to show a sea that was a maelstrom of torment, with white smoke-like spray partly obscuring the surface, streaks of spume trailing out in the direction of the wind and huge waves with breaking crests bearing down upon our starboard side.

From the bridge I watched a bergy bit as large as a two-storey house loom up 50 metres off to port, dipping and plunging amongst the spindrift. Frightening in its menace, it wallowed clumsily along, heaving one shoulder out of the water as the swell passed under it, then slumping drunkenly down again until almost submerged, only to rear up next minute with torrents of water pouring down its sides and being whipped away by the gale into misty smoke. With engines working furiously, the captain laboured to edge the ship past this danger.

At another time we were lifted high by a wave and our bows crashed down on the hard remnant of an old floe with a force that hurled splintered ice all over the decks and shook the ship violently from stem to stern. Fortunately, the stem of the ship was solid forged steel.

At about 0500 hours, on a visit to the bridge, I found that the seat and back of the settee in the chartroom had come adrift and were lying on the floor. A sailor came up and began to fit them back into their places. 'They're safer where they are,' I told him, but I don't think he understood English because he went ahead and completed the job. Some time later I was standing at the chart table when the ship heeled over violently and the two heavy pieces of timber-framed furniture flew through the air at knee-height and smashed into the doors of the cupboard beneath the echo sounder, missing me by a scant 18 inches. If I had been at the sounder or moving towards the door I would have collected two broken legs.

On one occasion, about 0700 hours, with the wind speed around 90 knots, we dipped our port side so far under that the portholes of the saloon looked out into green water. On another, the porthole of my cabin on the boat deck was completely submerged.

At about 0900 hours I went below and had some biscuits. There was no cooking being done and every cabin and store was a shambles. Nothing stays put when subjected to the violent angular accelerations of a ship being hurled about through an arc of 120°.

The captain never left the bridge. He had had a couple of blankets and a pillow brought up in case the opportunity rose for him to relax a little on the settee, but this was not to be for some time yet.

The lifeboats on the afterdeck began to move in their cradles in an attempt to follow the aircraft over the side, so the first mate took a work party of sailors out onto the heaving deck and roped them down. They then threw overboard a number of drums of aviation fuel that had moved in their lashings and were blocking the scuppers.

For the rest of the morning, as there was nothing I could do, I lay on my bunk or sat looking out at the storm. Of the men in the forecastle accommodation, only two essayed the dangerous trip over the exposed foredeck to the saloon. My lieutenant, Dick Thompson, and later Arthur Gwynn, ran the gauntlet, hanging grimly onto the safety ropes that had been strung along the hatches of the holds.

At noon I ate a few sardines from a can and looked out a saloon porthole to see a giant, creaming breaker tower up over the ship and come thundering down on the starboard quarter. It felt as though an iceberg had struck the ship. Fortunately, the main force of the impact was taken by the forecastle bulwarks and our motor boat, mounted across the forward hatch, escaped major damage.

Soon afterwards we drifted into an area of brash ice. Apparently, what had previously been pack ice had been pulverised by the waves into a consistency like concrete mix. This ice, composed of pieces varying in size from a few inches to a few feet in diameter, floated in a layer about four feet deep over the surface and completely smothered any breaking of the waves. However, the giant swells persisted and provided a most remarkable spectacle. At one moment, from the deck, we appeared to be in a white gully with sides sloping upward on either side of the ship to a height of 10 or 15 metres; at the next, we looked out from the crest over an undulating expanse of such gullies. As far as we could see stretched a vast, snow-covered countryside, in which ridges and valleys moved alternately up and down, slowly and majestically, like the horses on a merry-go-round.

Here and there in the 'concrete mix' floated bigger pieces of ice, remnants of ice floes, five or six feet thick and up to 20 feet across. Against these the ship grated and bumped with horrible, menacing sounds. So far there had been no change in the direction of the wind and little diminution in its force.

I judged we were somewhere near the middle of Prydz Bay, but there was no way of finding out exactly where, for no sun or star shots were possible and the type of drift motion we were experiencing did not lend itself to dead reckoning. I began to wonder what might happen if we drifted right across to the ice cliffs on the other side of the bay.

At 1600 hours I checked the weather out on the deck and found the wind had moderated to about 60 knots. There was no longer the frightening list to port and the danger of capsizing had gone. Also, the breaking waves, damped down by the brash ice, no longer smashed over the ship. Although there was still a lot of movement, the motion of the ship had decreased enough to allow the cook to prepare dinner. At 1800 hours we had a simple meal produced by warming up cans—tinned soup and curry and rice, followed by tinned fruit.

Another potential hazard had appeared. The spray from waves breaking over the ship and from blown spindrift had frozen and the decks and rigging were coated with a thick layer of ice. The weight of this extra load had further decreased the stability of the ship, and the captain had directed the crew on deck to chip off the ice. Fortunately since entering the brash ice, there had been no further accumulation and, as the temperature was not low (about 26°F), much of the ice on the rigging was now melting and dropping off.)

During the whole of this period I had not taken a single photograph! I can only say that, under the terrifying circumstances, there seemed so little prospect of our surviving that it appeared pointless to take photos. To my knowledge no-one else had taken any either.

After dinner, from the bridge, I looked out again on the grand but messy scene. No sea was visible, only the thick pulverised ice, alternating mountains high and valleys low, with large heavy floes heaving up and down in it and the ship bumping horribly against the bigger pieces. But out main anxieties had been relieved: the wind was now slowly moderating and it was possible to head the ship into the wind and resume steering. We were safe from the violent rolls of the broached ship that had threatened to capsize in the open sea and from the danger of giant breakers dashing bergy bits against our hull; and now that we could make way and steer, the iceberg hazard had considerably lessened and any danger of drifting into a lee shore on the opposite side of Prydz Bay had disappeared.

I went to bed at 2100 hours. The crew spent the night freeing the frozen pipes that had prevented the captain at the outset of the storm from ballasting the ship with sea water. I discovered, too, that frozen pipes had also prevented the transfer of fuel oil from one tank to another, for in each oil tank there is a certain amount of water, mainly from condensation. This, being heavier that the oil, collects at the bottom of the tank and, if frozen, blocks the outlets. Now, with sea water in the empty fuel tanks and topping up the after-peak freshwater tank, we carried over 100 tons of ballast, with noticeable improvement in draft and stability.

At 0300 hours on Saturday 6th, I rose and had a look around. The wind was still blowing over 50 knots, the second mate was on watch and the captain was resting, wrapped in blankets, on the settee in the chartroom. Floodlights illuminated the ice along the sides of the ship and a searchlight was trained full ahead. All was serene, so I went back to bed. I rose at 0800 hours after a good sleep. The wind was still north-east and gusty, but its general speed was down to around 20 knots. Visibility was poor with mist and snow showers. I went down to a normal breakfast.

Morale had improved immensely amongst members of the crew and the expeditioners. (The previous night it had been rock-bottom, for no-one had given us much chance of surviving.) Nevertheless, a number of the younger lads in the crew were saying now, 'Never again! You can keep the Antarctic! All I want is a shore job from now on.' The normal watch routine on the bridge was resumed and the captain at last retired to his bunk after 40 hours continuously on the bridge.

By lunchtime we were proceeding at 2-3 knots in a direction north-north-east with a 20-knot head wind, a 26°F temperature and visibility poor because of frozen mist. The ship was upright and steering comfortably, but progress was slow because of the several feet of brash ice that covered the surface of the sea and reduced our speed because of its friction on the hull. It was like trying to sail through glue. I spent most of the day writing an account of our Vestfold Hills observations and I sent a dispatch to the Antarctic Division. I kept it low key, for I did not want to alarm our next of kin.

I got to bed about 2130 hours, wondering how much pack ice remained between us and the open sea. I could hardly hope that the whole belt of pack would have been pulverised and dispersed. In any case, lower March temperatures could freeze solid the pulverised brash and trap any ship trying to proceed through it. We were safe after a period of gravest anxiety, but we were not yet out of difficulty.

* * *

This account was published previously as Chapter 12 in my book *Antarctic Odyssey* (Heinemann 1983).

2

MOUNT HENDERSON, MACROBERTSON LAND

It was Thursday, 24 February 1955. Mawson Station had been established the year before and now I and my men had returned for the relief of the pioneer party that had just spent the first year there.

For several stormy days we had been busy unloading and building new huts, but this particular day dawned sunny, cloudless and promising to be fine. It was windy, 20-30 knots, but later it calmed down after about 1000 hours.

We decided to take a party in to Mount Henderson, which was the closest mountain in MacRobertson Land to Mawson Station. Dovers, the officer-in-charge the year before, had visited the mountain and had pushed on beyond it to inland nunataks* giving a view of the Prince Charles Mountains in the distance. The idea was that Dovers should take the new officer-in-charge and several of the new party to show them the route from Mawson to Henderson, which was rather difficult because of the many crevassed areas on the way. Dovers had flagged a route by erecting cane markers with black flags.

We got away at about 0930 that morning. There was one Weasel (an over-snow tracked vehicle powered by a six-cylinder Studebaker engine), with Dovers driving, containing Béchervaise, Somers and Lowe. There was a second Weasel, me driving, containing Gowlett, Lacey and Shaw. There was also one Army DUKW (amphibious six-wheeled vehicle) with Lieutenant Hall driving and carrying mechanic Power and journalist Hungerford. Taking the DUKW was an experiment—no DUKW had ever before driven up onto the plateau.

Each Weasel was towing a cargo sledge with four drums of petrol and Dovers had another sledge behind his with a spare Weasel track. Up

* There is a glossary of ice and polar terms on page 195

the first steep, iced slope from the station the Weasels could not grip on the mirror-smooth ice—the result of a warm sunny afternoon followed by the usual very cold night. So the DUKW, fitted with chains, had to come back and tow each Weasel. After that, things were fairly straightforward, except that my Weasel kept heating up and this forced us to run in second gear, low ratio, at about six to eight miles per hour. The DUKW could have raced ahead at 15 miles per hour and its crew felt frustrated by the slow progress made by the Weasels.

We had a radio schedule when not far from Mawson at 1130 hours but my radio would not transmit. Dovers had put in about six black flags along the route to Henderson.

At noon we reached a depot that Dovers had established early in his year on a rocky knoll a little to the north-east of Mount Henderson. We noted from the top of the knoll that the DUKW, which we had left behind at the radio schedule place, had bogged. Dovers returned in his Weasel to help but I couldn't start mine and Gowlett found that the battery lead to the starting motor had become disconnected. It took him some time to fix the fairly inaccessible fault. In the meantime, the DUKW came in, having winched itself out of the soft snow in which it was stranded. Dovers and Béchervaise immediately set off to examine the southern route through a heavily crevassed area which lay four or five miles further inland. The rest of us took photos, had lunch and unloaded and stacked petrol drums.

After this I went for a walk. I climbed over a rocky knoll on the south-east side of the depot and saw Mount Henderson rising boldly before me out of a sheet of ice about a mile long separating it from the knoll. I saw that to get from the knoll to the ice and from the ice to Henderson would be fairly simple. Simple, also, appeared to be 80 per cent of the climb to the summit of Henderson. The last steep pinch looked reasonable but I could not see at that distance what difficulties it might present. I decided that I would like to climb it, but I was worried by the lateness of the day. As I returned, I saw Dovers' Weasel pass on its way back to the depot. When I reached the depot I immediately asked Dovers could he wait until 2000 hours while we attempted the climb. He agreed. He did not wish to come, nor did most of the others, so the final climbing party comprised myself, Béchervaise, Shaw and Hall.

It was 1620 hours when we left and we were soon on the bare ice, where we fitted crampons to our boots. The walk across one mile of this ice to the foot of Mount Henderson was pleasant, although the wind blowing on our right cheeks was extremely cold. At the foot of the mountain we left our crampons and ice axes and started the scramble up about 800 feet of scree and boulders. Hall, in his usual enthusiastic fashion, streaked out ahead and, although I normally don't care whether I am first or last in any party, I resolved I would beat him to the top of the scree.

In his first vigorous dash Hall had expended a lot of energy and he continued to alternate fast bursts with short rests. As he tired, the bursts became shorter and the rests longer. I had established a steady pace that I believed I could maintain without resting and I passed him about 30 metres from the top of the slope of scree, as he stopped to puff and pant. Much to his disgust I led him up to the main rock face by 15 metres.

In my haste I chose a rather bad spot. Shaw, in his usual methodical, careful fashion, found an easy way up the first rock pitch, only some tens of metres further around the face. He had to reach a hand to help me up over my difficult pitch. After a bit more scrambling we faced the main final 50-foot pitch of rock to the top. There were two alternative routes and, after discussing it with Peter Shaw, I decided on one. Tony then rushed in to attack it and I spoke to him to point out that he had no experience in rock climbing, we were not roped and he should be the last person to attempt to lead what was a fairly difficult climb. We could not afford to have an accident at this place at this time.

I started off and soon met trouble. I could have gone on while trying two alternatives but I was worried that if I slipped, not being roped, I might be injured and cause embarrassing delays to the party. Shaw, again, had used his brains; he had scouted around and found a third possibility around behind the main face and had scrambled up that with no great trouble.

As there was no point now in risking a tricky climb unaided, I asked Peter to pass down a climbing rope. Safeguarded in this fashion I successfully climbed this particular pinch. Tony, impatient as ever, didn't wait for the rope. Being particularly long in the legs he was able to stretch over some awkward distances and he got to the top unaided. Béchervaise followed us, using the rope like any sensible well-trained mountaineer miles from help on a difficult face. We were up by 1715 hours, just under two hours after leaving the depot.

On top it was bitterly cold. Shaw measured it as being 2°F (-30°C) The wind was blowing 30 miles an hour and altogether it was no place to hang around. The space on top was very limited, an area about five metres by four metres, and we clustered there taking photos. My Leica camera film snapped because of the cold so, when my Rolleicord black-and-white film was finished, I opened it and put in a Ferania colour film. To change a film was a silly idea on my part, for I had to take my gloves off to do it and my fingers froze, sticking to the metal. I unstuck them and replaced my gloves but my hands did not warm up. I was now shivering and stiff after sweating during my scramble up the scree so, as it was by then 1900 hours, we decided to hurry down.

Béchervaise made a loop of a piece of our nylon rope and we wedged it in a crevice with a rock to make an abseil sling belay. Shaw belayed me down and I found that going down was much more difficult than going up. I had to remove my outer wind-proof leather mittens. The wind then whistled through my woollen mittens and froze my hands, which were pressed hard against the cold rock for support. When I reached the bottom of the face and unfastened the rope and sent it back up to the others, I removed my mittens and found the tips of all four fingers of each hand were white and frozen. (If this is allowed to continue for long, one has frost-bitten fingers.) I managed to thaw them in my mouth and under my armpits but, as the circulation began to return, the pain was agonisingly intense and I danced around, waving my arms, hoping that the agony would not be too long prolonged. If you have never had frozen fingers and have never experienced the recirculation pain, you will not understand how agonising it can be.

Béchervaise belayed Hall and then Shaw. He followed them down on a double rope, and recovered the rope afterwards.

From there we had an easy scramble down the scree to the ice. On the way, quite near the top, Shaw and I had found a rock-like substance that looked like cream which had flowed down into a hollow and then frozen. We broke some off and brought it back. Later we found that it was frozen excrement from snow petrels.

The view from the top of Mount Henderson had been magnificent. Ten miles away was the ocean and the islands beyond Mawson, with icebergs sparkling in the blue water. To the east there was nothing but the snow plateau. Westward, however, was a fascinating vista. First there was the coast, winding away mistily into the distance, where a further view was cut, off by a large glacier. Inland, the Framnes Ranges—Masson, David and Casey—with their numerous abrupt peaks rose stark and black from the blue ice. Some peaks were beautifully symmetrical and pointed like paper caps. The sharpness of their points is what I think is so startlingly different, used as we are to Australia's old worn mountains.

There were interesting glacier crevasses streaming out from a bergschrund near where we first had donned our crampons. We saw one snow petrel high over Mount Henderson. There were many lichens, including one snow-white flocculent type I had not seen previously, growing only on permanent rock at one level.

We regained our crampons and ice axes at the edge of the blue ice and walked the long way back around the edge of the knoll, staying on the ice. The ice was criss-crossed with narrow crevasses which had iced over during thaw periods to give transparent, blue, glass-like cracks in the surrounding pale white surface. As we walked, the ice surface beneath us cracked with sounds like toy caps in a child's pistol.

We were now quite tired and were delighted to find Dovers come to meet us in a Weasel. We climbed gratefully on board at 2015 hours and he drove us the last mile to the depot. There we had a cup of hot water and sugar, some chocolate and a biscuit each and then we packed and set off in the vehicles back towards Mawson.

The DUKW went first. I told Hall to head for home and to come back for us if we had not arrived an hour later than he. I left with my Weasel from the depot at 2050 hours. Dovers waited back for a radio schedule at 2100. The wind was now blowing hard and the temperature at the depot was 5°F. I followed our outward tracks but, about two miles from the depot I lost them and, as the heavily crevassed terrain seemed much rougher than on the outgoing journey, I turned further west and had just begun to wonder where the real track was when my engine stopped. There was a smell of rubber and some smoke and we found the main battery lead to the starter had fouled the exhaust and burnt through, shorting onto the frame. The resulting surge of current had fused the earth lead of plaited copper from the battery to the frame. Gowlett joined the earth strap with copper radio aerial wire and found that, luckily, the main leads in the burnt rubber insulation were still okay. So we got going again just as we saw Dovers coming behind us. He was travelling in nearer the mountain, so we headed back and picked up the tracks again.

We found that Hall in the DUKW had also missed the track. He had turned west to try to cross a melt-water channel but his rear wheels had become bogged. All went well with our Weasels until about four miles from Mawson, where we came upon the bogged DUKW, which had sunk down to the undercarriage. The men had been unable to winch the vehicle out. It was now dark and blowing 40 miles an hour, with a temperature of 10°F. The ice surface was treacherously slippery and we were all sliding and falling. We used axes to dig away ice from under the wheels and frame of the DUKW; then we put jacks under the rear section, sat all our men on the front of the DUKW and it managed to pull itself out forwards under its own power. All this took about one and a half hours and it was about 2230 when we got moving again. I led the party in my Weasel to try and find the track with my lights, which were better than the others'. Then came the three sledges, then Dovers holding the other end of the sledges, then Hall's DUKW. It was most difficult following the morning's outward tracks, as already most of them had evaporated or melted off. Sometimes one could see crystal indentations and sometimes patches of packed drift where the tracks had been, but it was quite a strain in the dim headlights.

There had not been enough flags set on the outward journey to give us guidance. Eventually, I lost the tracks altogether, so unhitched and scouted around singly and unsuccessfully for 20 minutes without finding them. After a chat with Dovers I decided to camp the night. We divided up—six in my Weasel, five in Dovers'. As the DUKW would have been too cold, the DUKW men joined us in the Weasels. We had one biscuit and half-a-block of chocolate each and curled up in our sleeping bags at about midnight with the wind buffeting the sides of our vehicles and the drift swirling around at knee level outside.

I worried most of the night about the weather, for it had blackened up and was obviously snowing to the north-west. If visibility were to turn bad or if there were a blizzard we would have to stay put for perhaps two or three days and we had no food and no stores. We had left our sledge boxes containing these to stock the depot at Mount Henderson. We were warm and comfortable, but cramped into a five-foot length so that, when my leg muscles began to ache, I finally achieved relief by stretching them up into a corner of the roof of the cabin.

I guessed we were four to five kilometres from Mawson but a very tricky four or five kilometres.

I rose at 0525 hours the next morning and drove my Weasel to wake the others It was very nice not having to get out in the cold. We hitched up and set off at 0600 hours in good visibility but in a 60-70 kilometre per hour wind. I sent the DUKW on ahead. All went well; tracks were impossible to see in many places, but we managed until we came to the last ice slopes leading down to Mawson Station. There my Weasel was overrun by the sledges which we were towing. Bob was unable to hold them back with his Weasel. The weight of the sliding sledges swung my Weasel around until I was facing uphill; then, with my engine racing and tracks scrabbling, I was just barely able to hold my position on the ice.

We unhitched the sledges and drove the Weasels separately down this treacherous ice slope to the station. From there we sent Hall back with the DUKW to collect the sledges.

We arrived at the station at 0700 hours after a grand experience.

Incidentally, the wind continued to blow at around 60 miles an hour for most of the morning, but at lunchtime it died away and we were able to start normal station work and unloading from the ship in the afternoon. I continued with this work and returned after dinner to the ship, where I had to write my diary, fix my films and cameras and do various odd jobs. I finally fell into bed, exhausted, at about 2230 hours.

3

THE DRIVE FROM OMEO TO MITTA MITTA

In 1932 I was studying Part I Science at Melbourne University while enrolled as a non-resident student at Melbourne Teachers' College. A fellow student, Bill Boswell, who was doing Part II Science, became a close friend. At the end of the university year, I persuaded him to come with me on a walk across the Bogong High Plains in the Australian Alps.

This undertaking was an exceedingly primitive one. I had had camping and bushwalking experience and mountain-climbing in the Grampians, but I had never collected any specialised camping gear. I did not own a tent. Sleeping bags were almost unheard of—I certainly didn't have one as I'd always used blankets when I was camping. I didn't have a rucksack. (The rucksack in those days was a European product that I had heard of but never seen.)

The plan was to travel by train and bus to Harrietville, at the foot of Mount Hotham and Mount Feathertop in north-eastern Victoria. There we hoped to hire a pack horse, but we were unable to persuade any of the local farmers to provide one.

So, at the pub in Harrietville, we discarded everything not absolutely essential and despatched it back to Melbourne. Then we tried to arrange some means of carrying our load. Looking back, it was quite absurd. We just adopted the methods of the swagmen of the depression; we rolled our blankets and bent them into a 'U' shape, placed the 'U' part over one shoulder then tied the two ends down on the opposite hip. Then we slung over the other shoulder various bits and pieces—a pot and a frying pan and mugs and plates. We took food for one week, but no tent because we planned to stay in cattlemen's huts or camp out. I hung onto my rifle.

Next morning, we rose and left Harrietville at 0400 hours. We had no maps, so we just set off along the road to Mount Hotham.

This Alpine road rises from Harrietville to the summit of Mount Hotham and then gradually descends across the high country to reach Omeo in the far distance to the east. On a saddle just below the summit of Mount Hotham was the Hotham Heights Chalet. From Harrietville to the chalet was roughly 33 kilometres.

Our objective that day was Dibbin's Hut, which lay in a valley 300 metres below the level of the chalet. We arrived there at sundown, having walked 41.5 kilometres, the first 32 of which rose 900 metres uphill. The last eight kilometres included a steep descent of 300 metres. We were quite exhausted, but had to clean up the hut and cook dinner.

Dibbin's Hut remains in my memory as a beautiful example of the romantic idea of a rustler's hideout. It was a log cabin nestling in a very narrow, gorge-like valley with a crystal-clear stream bubbling along over rocks past its front door and a 300-metre rise of mountain each side. It was enshrouded in snow-gum trees and, in front, a sward of green grass and wildflowers led to the creek. It had a rough dirt floor, a large open fireplace, a few shelves nailed to the walls and, on one side, two layers of bunks made by stretching hessian bags between horizontal logs. It was infested by rats which ran around and crawled over us all night and kept us awake. Next morning we rose early, climbed out of the valley onto the Bogong High Plains and set out for the next cattlemen's hut.

Over the next three days we stayed in two huts that I can remember, Roper's Hut and Fitzgerald's Hut. We traversed the Bogong High Plains and descended at the far end into the valley where we hoped a little settlement called Glen Wills lay. Glen Wills had been a mining town of some substance but most of the mines had closed until, at that time, only one was still running—a small mine called the Maud and Yellow Girl Mine. We had expected to find a post office and a store and a few houses, but no. There was only a boarding house, the mine and, further down in the valley, a few farm houses. No store, no post office.

Mitta Mitta was 65 kilometres further on but, without food, there was no way we could attempt to walk that distance. We tried unsuccessfully to buy food from the boarding house. So that night we rolled up in our blankets in the open, beside a fire. Although I had shot a rabbit, we dined on some semi-putrid ham which we fried.

We were in a dilemma, marooned in this isolated spot without enough food to travel elsewhere. Fortunately, we learnt from the boarding house next day that a chap about to drive to Mitta Mitta might give us a lift.

We duly met Charlie, who agreed to take us in his old National model Chevrolet. It had a canvas hood, but no side curtains, and was of mid-1920 vintage. With him was his young sister, aged about 13 or 14. Charlie must have been about 19.

We set out with Bill and me in the back and our gear on the floor. In my hand I held my rifle, which I had used to shoot the rabbit the night before.

The road from Glen Wills to Mitta Mitta is a typical mountain road which, even today, is quite frightening, although it is sealed and wide enough for two cars to pass. In those days it was a narrow gravel road only wide enough for one car; in many places a cliff rose from the road on one side while, on the other, a steep incline fell away for 60 to 90 metres into a torrent of river. The road was just a collection of sharp bends, one after the other, around which one crept, tooting the horn to warn any oncoming car. If another car were met, one car would have to back for some distance to find a place wide enough for the two to pass. Fortunately, we didn't encounter any. However, as we proceeded, I became more and more apprehensive—a feeling of insecurity arising from the steep slope to the river below.

I remember planning what I would do if we had an accident. I was on the side closest to the cliff; Bill was on the other side, looking down over the drop into the river. I rehearsed how to get out of the car in a hurry, planning where to put my feet and my hands. I decided not to bother about opening the door for I would probably not have time.

About half-way to Mitta Mitta, we rounded one very sharp bend and I suddenly noticed that we were not coming out of our turn—we were heading for the edge. Just as I was wondering what the hell was happening, the driver called 'Jump!'—and jump I did. Quite subconsciously I must have put my feet and hands in the various planned positions and hurled myself out over the top of the door onto the road, landing on all fours and sustaining gravel-rash on hands and knees. I was still clutching my rifle! But I was safe and I had a surge of exhilaration as I thought, 'My God, I'm alive and the others are probably all dead. But I'm here, I'm alive.' I went to the side of the road and peered over, expecting to see the car smashed to pieces in the river below. To my astonishment I saw that there was a substantial gum tree sticking out about five metres below the road. Into this, the only tree on the slope, the car had wedged its radiator. In the car the three passengers were sitting bolt upright, white as ghosts. They were completely unharmed. I was the only one who had suffered any injury.

They clambered out of the car and climbed back onto the road. The only thing to do was to walk to Mitta Mitta, a distance of about 32 kilometres. So, in the late afternoon, we set off, with me carrying the rifle and the rabbit.

About 2100 hours that night, in darkness, we rounded a bend and saw a campfire on a small flat. There we found a young hawker, about 18, with a horse and gig, who had camped for the night. In those depression days work was hard to get and this chap was making a frugal living travelling around country districts selling odds and ends to various farmhouses. He agreed to harness his horse and drive Charlie and his sister to Mitta Mitta. Bill and I camped beside the hawker's fire. He returned late that night, by which time we had roasted and eaten the rabbit.

The hawker assumed that Bill and I were hoboes, because we had a week's growth of beard and were dirty, dishevelled and unkempt. So he gave us a run-down on all the farmhouses and property owners in the district, noting those who were kind, those who would give you a feed, those who would give you accommodation and those who would turn the dogs onto you.

Next morning, Bill and I began to walk to Mitta Mitta, but soon a motor truck came up the road driven by 'Tiger' Lyons, who owned the pub at Mitta Mitta. With him was Charlie—they were going up to rescue the car. The truck was able to tow the car up onto the road and it was not so badly damaged that it could not drive. The main problem was that the steering was in a strange state. A 'ball-race' had disintegrated, but Charlie had found that he could spin the wheel one or two revolutions to the right and then it would grip and turn the wheels. Then he'd have to spin it one or two revolutions back to the left before it would grip and turn the wheels the other way. With a bit of practice he was able to steer the car down to Mitta Mitta.

Bill and I went with Tiger in the truck. I found that trip more frightening than the trip in the Chevrolet, as Tiger drove furiously around all the bends without worrying whether any other car might be approaching; but we arrived safely. Later we drove in the Chevrolet again from Mitta Mitta to Tallangatta—a distance of some 60-odd kilometres. That journey, too, was rather hair-raising as Charlie spun the steering wheel back and forth around the sharp bends along the hillsides.

4

CROSSING THE PETERSEN BANK

The year was 1956. For the first time, there was no Heard Island station to relieve and I was free to use ship-charter time for exploration.

For the coming summer season I had drawn up a program designed to map as much as possible of the unknown sections of Australian Antarctic Territory between Adélie Land and the Shackleton Ice Shelf, then to revisit Prydz Bay and later, after relieving Mawson, to push on to Enderby Land.

The first part of the plan worked well. We completed valuable exploration in the vicinity of Lewis Island and Chick Island (where, in later years, we were to operate automatic weather stations). Our next objective was Vincennes Bay and the Windmill Islands, where men of the US expedition Operation Windmill had made brief landings in 1948.

During the afternoon of 18 January 1956, *Kista Dan* was approximately in the position 65°S, 111°E and heading south-west along a fairly open lead with 7/10 pack ice on each side.

By 1900 hours the Antarctic Plateau was visible in the distance. The floes were all small to medium in size, mainly low, with moderate rafting, with high floes here and there a couple of metres above the water, with a metre of fresh snow on top. Most of the icebergs were small glacier icebergs, but there were a few tabular ones and several of the domed variety. The ocean bottom was planed perfectly flat at a depth of 150 metres.

I spent two hours during the evening in the crow's-nest with Captain Hans Christian Petersen. Ahead we could see some rock outcrops that appeared to be part of the mainland, but they were not marked as such by the Americans on their 1948 sketch map, so we headed directly towards them. Shortly after 2300 hours the ship stopped for the night.

We moved on at 0545 on 19 January. I gave the new Mawson OIC (Officer in Command), Bill Bewsher, a list of activities his men were to carry out if we were to achieve a landing; my deputy, Dick Thompson, would supervise the logistical requirements of the operation. At 0756 hours the ship dropped anchor off a small island close to the ice cliffs of the mainland. This and other nearby islands constituted the rocky outcrops we had seen. The US had named the group the Balaena Islands after the whale factory ship that first sketched part of this coast.

The second mate landed Thompson, Bewsher and me by motor boat. On a second trip he brought ashore Kirkby (surveyor), McGregor (geomagnetician), Ayres (NZ observer), Bunt (biologist) and Dowie (medical officer) and they carted all their gear to the top of the island. On a third trip some of the remaining men came ashore and helped to build a cairn on the summit.

There were six islands in the group, of which ours was the largest. From the summit the scene was unforgettable—the smooth sea, sprinkled with floes, bergy bits and icebergs; the scarlet ship motionless against the sheer ice cliffs of the continent which sloped back steeply towards the hinterland; the warm, chocolate-brown rock, the curious penguins and, above all, the placid peace and calm which only a tranquil day in Antarctica can give. The squawks of penguins carried clearly from far-off floes, and the little chugging motor boat cut through the floating chips of ice towards the landing place, leaving a fantail of spreading ripples. Off the northern edge of the island an old bergy bit, newly capsized and glazed with fresh ice, gleamed a burnished blue in the slanting rays of the afternoon sun.

Ours was a typical 'hit and run' landing in which each of our specialists extracted the maximum amount of information in the shortest possible time—cartography, geology, geomagnetism, botany and zoology. Our observations completed, we sailed at 1640 hours.

Our immediate destination was a large ice floe, a few kilometres to the east, which we had passed earlier and which we proposed to use as an ice airfield for a flight by the Beaver. Damaged floats precluded launching it from the sea. However, there was a moment of excitement as we headed towards the floe, for the sea floor rose steeply from a depth of 60 metres to one of eight metres before the ship could be stopped and backed off.

Kista Dan tied up at the ice floe airfield at 2000 hours and the Beaver took off on skis at 2030 for a flight south-westwards to the Windmill Islands. Peter Clemence was the pilot.

We stayed put overnight and, next morning, Leckie gave John Seaton a couple of practice take-offs on skis before he and Leckie set out on a run with the intention of photographing the coast through Cape Poinsett to the Totten Glacier. However, they were turned back at the Cape by heavy cloud, so I directed them by radio to fly SSW to repeat Clemence's run of the Windmill Islands, which had been carried out in poor light.

They landed at 1500 hours, we hoisted the plane on board and immediately proceeded westwards, aiming to follow the coast around to the Windmill Islands in Vincennes Bay. However, just after passing the Balaena Islands, Hans came to me and said he could not go further. The depth was only 40 metres, a lot of the large floes were grounded and a two-knot tidal current was making steering difficult at dead-slow speed. He proposed returning to open water 30 kilometres to the north before proceeding westward again. I said 'Why not go three or four kilometres further north and try again? The water might be deep enough there.' I was concerned at the loss of valuable time involved in having to go 30 kilometres north and, later, 30 kilometres south again. Hans then questioned the RAAF men about the ice conditions, but they told him that the pack ice became thicker further north—so he agreed to try my proposal.

As time wore on after dinner I noticed Hans was edging further and further out from the coast towards the accumulation of icebergs that ran down to the mainland to the west of us. They were obviously stranded on a bank. By keeping close to the larger bergs he was ensuring enough depth of water for his ship. However, the pack was denser near the bergs, so he cautiously worked his way southwards again at dead slow speed until only about 10 kilometres from where the iceberg bank joined the continent. Here there was quite a bit of open water between the floes and the icebergs were not as densely packed.

At 2030 hours I climbed the ladder to join Hans in the crow's-nest. The next few hours were some of the most exciting and tense moments in all my sailing experience. We were to attempt a short cut to Vincennes Bay directly across the shallow bank.

In the dim half-light of the Antarctic night, in calm clear weather, the *Kista Dan* weaved her way slowly forward as we peered ahead to select routes and listened apprehensively to the soundings being relayed by the first mate over the intercom from the bridge. The best depth indications we had were the icebergs, bergy bits and floes, for if these were tilted it showed that they were aground, and we could make a rough guess at the depth of water according to their size.

The old hands on board realised the drama of the situation and stayed up to watch, but most of the 1956 Mawson party had gone to bed. Kirkby stood beside the first mate plotting depths on the echo sounder. Thompson stood alternatively on the wing of the bridge, peering anxiously at the tilted floes, or inside the bridge, watching the sounder and talking with Kirkby and the mate. A few other men had sought vantage points on the wings of the bridge or the boat deck.

From the crow's-nest, Hans and I saw ahead a large area of open, ice-free water, but it was not till we reached it that we realised it was clear of bergy bits and floes because it was too shallow for them to pass over it! Yet now we were very close to our objective—Vincennes Bay—and retreating to the north to round the bank would mean the loss of at least one day. To Hans, more importantly, it would mean capitulating to the sea and ice, and he became as keen as I was to persevere.

So on we went, avoiding the open water, weaving amongst the floes and choosing the large, level ones to sidle past. We saw others, barely 50 metres on either side of the ship, tilted awkwardly and stranded in just a few metres of water.

It was quite eerie up the mast and, in the calm conditions, silent except for the soft throb of the dead-slow engines. I was cold and stiff and tired—there was no seat for me in the crow's-nest—but I could not tear myself away from the fascination of this contest between the captain and the elements. Finally, at 0016 hours, Hans decided the light was too poor and anchored for the night in 12 metres of water!

Next morning, 21 January, the ship started again at 0625 and we repeated our previous performance (Hans conning the ship from the mast, edging her forward slowly, swinging her bows this way and that, listening to the depth calls from the bridge and choosing first one and then another untilted ice floe to bring the ship close by.

At last, around 0800 hours, the water began to deepen and we were able to breathe more normally. I congratulated Hans and he grinned appreciatively, his grim concentration giving way to an almost boyish delight and satisfaction. We had crossed the Petersen Bank (as we were later to name it) and most of the crossing had been done in the dim Antarctic twilight!

The feat was an exercise in courage and seamanship by Hans Christian Petersen, and it is unlikely ever to be repeated. It should be pointed out, in this context, that this happened before the IGY (International Geophysical Year). There were no ships of other nations anywhere in Antarctica that could have come to our assistance had we run into trouble.

Over the following days we were to make several new landings in Vincennes Bay and to familiarise ourselves with one of Antarctica's large areas of coastal ice-free rock. We took astrofixes, collected scientific information and made aerial photographic runs from Cape Poinsett in the east to the Bunger Hills in the west. On the way north again we plotted the western edge of the Petersen Bank.

The Windmill Islands seemed to me an ideal location for a permanent Antarctic station, the only place in Greater Antarctica that could match the areas of Mawson and Davis for such a purpose. I was not to know, then, that circumstances would ultimately make it possible for us to establish our third mainland station there (Casey Station).

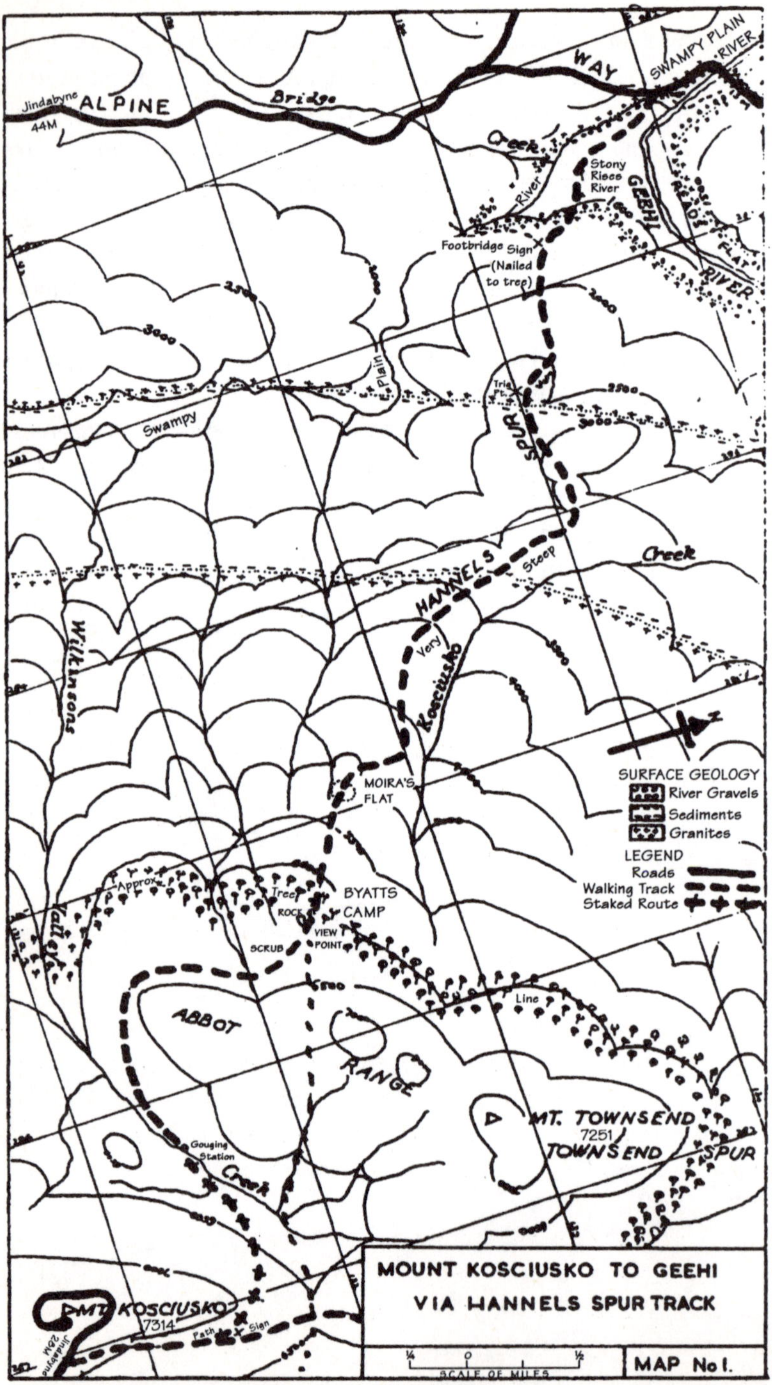
Jindabyne
44M
ALPINE
Bridge
WAY
SWAMPY PLAIN RIVER
Creek
River
Stony Rises River
GEEHI
READS FLAT
RIVER
Footbridge Sign (Nailed to tree)
Plain
Swampy
Trig. Pt.
SPUR
HANNELS
Steep
Creek
Very
Kosciusko
Wilkinsons
MOIRA'S FLAT
SURFACE GEOLOGY
River Gravels
Sediments
Granites
LEGEND
Roads
Walking Track
Staked Route
Approx
Valleys
Tree
ROCK
BYATTS CAMP
SCRUB
VIEW POINT
Line
ABBOT
RANGE
MT. TOWNSEND
7251
TOWNSEND SPUR
Gouging Station
Creek
MT KOSCIUSKO
7314
Path
Sign
MOUNT KOSCIUSKO TO GEEHI
VIA HANNELS SPUR TRACK
SCALE OF MILES
MAP No 1.

5

Mount Kosciusko The First Trip

Mount Kosciusko, at 2250 metres, is the highest mountain in Australia. Its western faces lunge down 2000 metres into the Swampy Plain River at Geehi. This river and the Geehi River join there to form the headwaters of the great Murray River.

In 1936 I was teaching at Elwood Central School and my elder brother Geoff was teaching at a small country 'rural school'. Geoff persuaded me to accompany him on a trip which aimed at reaching the top of Mount Kosciusko by ascending Hannels Spur from Geehi. Geoff had enquired, by letter, of the Corryong police about access to Mount Kosciusko from Geehi. They had replied saying there was little to be gained by wandering around the gullies of Kosciusko and bluntly advised him to stay home!

When school broke up for the May vacation, we loaded up Geoff's T-Model Ford and drove to Seymour, where we camped. Next day we drove to Tallangatta and stayed overnight with our auntie. On the Sunday we drove to Corryong on the Murray River and then on about 28 kilometres to Khancoban.

Just beyond Khancoban we met a drover called Drummond, who was pushing a herd of cattle across a ford of the Swampy Plain River. We sat and had lunch with him and he plied us with pint mugs of port from a demijohn. Half-drunk, we drove on about 15 kilometres to where a suspension bridge of fencing wire and six-inch planks led across the river to Waterfall Farm, which marked the end of the vehicle track from Corryong.

Geoff had written previously to the owner, Errol Scammel, asking could he hire us a horse and Scammel had agreed. We found him most cooperative and duly departed with a horse, complete with pack-saddle. We were quite ignorant of the route from Khancoban to Geehi, but

Scammel had given us brief instructions. It was simply a bridle track that started off climbing steadily, then ran along a long ridge and, finally, plunged down a very steep 300-metre drop to the Geehi River below. This incline was called the Geehi Wall, and very aptly.

We had told Scammel that, if we did not return in 14 days time, he should come looking for us.

I had ridden a horse as a small boy but had not had any experience with one since and I remember being a bit apprehensive as the horse picked its way along the ridge above the Geehi Wall, stepping over branches and logs and making traverses across steep side-slopes. But it was more frightening at the Geehi Wall, where recent rains had produced a lot of run-off and where, in places, the slope consisted of shallow curtains of rushing water flowing over gravel, the whole being quite slippery for horses. My horse wanted to turn its head and body sideways and, thinking that the horse probably knew better than I, I allowed it to do so, with the inevitable result that its feet slid from under it and the horse and I slid 40 metres downhill before we came to rest against a tree. From there on I led the horse down on foot. Geoff kept his horse's head pointing downhill and the horse finished up sliding down on its hindquarters for some 20 metres.

The track from the foot of the wall to the Geehi Hut, where we planned to have our headquarters, made about eight crossings of the Swampy Plain River. As we wound along the flats we would come, every mile or two, to another crossing. For the packhorse this was no great problem, but we had to wade through.

In general the water was between knee and thigh in depth, flowing swiftly, and the stones on the bottom were large and slippery. We used to strip from the waist down, but then put on our boots again to protect our feet while stumbling over the submerged stones. Our feet and legs would ache painfully in the icy water and we would jump around vigorously after crossing to restore the circulation.

We arrived in the late afternoon at the small grassy open valley where the Geehi Hut was situated, but it was on the opposite (eastern) side of the river. So, once again, we had to cross.

The hut was built of solid logs, with a huge fireplace and a chimney of galvanised iron at one end and a dozen sacks of rock salt (for cattle) inside the door. A bench table was attached to the wall on one side while, on the opposite wall, two bunks were fixed, one above the other.

Early next day Geoff and I began to search for Hannels Spur. We were not sure where it was. We had no map. Numerous spurs ran up from the Geehi Valley to the massif which had Mount Townsend (2231 metres) at one end and Mount Kosciusko (2250 metres) at the other. All we knew was that, on Hannels Spur, there was a track cut through the timber in 1930 by two cattlemen, Hannel and Pierce, who wished to take cattle up and down to the Kosciusko high areas in summer. However, it was so steep that it was never used for that purpose and gradually it became overgrown.

Leaving the Geehi Hut, we had to go back across the Swampy Plain River again to the west bank. Having crossed, we walked upstream, opposite a likely-looking spur, then we waded through the river again to the east bank and searched the foot of the spur for the track. No luck. For the rest of the day we tried spur after spur, even climbing high up on several, trying to intersect the track if it were there.

We had one bit of good luck. A couple of kilometres upstream from our hut we discovered a 'suspension bridge' over the river, consisting of two wires, one five feet above the other, slung across the river between two trees on opposite banks. One held onto the top wire with one's hands and shuffled one's feet along the bottom wire; in the middle the whole thing swayed dangerously, so one moment one was face-down looking a the water, the next moment one was face-up looking at the sky.

That night, defeated and exhausted, we retired to our hut for dinner and bed.

Next day we tried again and had the good luck to meet several cattlemen with packhorses on the track beside the river. They pointed out roughly where Hannels Spur was, but we climbed another couple before finding it. Back we went to spend another night in the hut, but this time our objective for the next day was known. We had turned the horse loose in a grassy area corralled off near the hut. It would be all right for the next couple of weeks.

On the third day we rose at about 4 o'clock, loaded ourselves with food, skis, sleeping bags, rucksacks—everything we needed for the following week—crossed the wire bridge, walked upstream for several kilometres, waded through the river four times to get our gear across and then began the ascent of the spur.

It was hard climbing—this spur is the longest, steepest approach to any mountain in Australia, rivalling even the Staircase Spur on Mount Bogong.

We stopped for a rest and lunch at an attractive area called Moiras Flat. Then, pushing on upwards through snow-gums and scrub, we reached the 'tree line' at about 1800 metres. From there on it was snowgrass, gradually fading out until, near the top of the Abbot Range, it was boulders and scree. About 150 metres below the summit of the Abbot Range we came into a field of giant boulders. This was quite an astonishing stretch of terrain. The boulders varied in size from a metre diameter to those the size of a small room and some as large as a house. We had to scramble around, through, under and between without being able to see where we were going, except to make sure we kept climbing. It was a relief to emerge from the boulders onto the top of the Abbot Range, where we had our first view, across a depression called Wilkinsons Valley, of the summit of Kosciusko. We had to descend about 60 metres into Wilkinsons Valley, then climb out on the other side. By this time it was late afternoon; we had been moving all day, except for a short rest for lunch, and we were exhausted. We had climbed roughly 1850 metres with our loads on our backs.

A sheep fence of wire stretched across the valley. The easiest way of getting through it was to lie down and wriggle under the bottom wire. Geoff went first and, when he lay down on the ground to wriggle under the wire, he was so exhausted he just stayed there and went to sleep. I kicked him up and forced him to go through and then I had the same experience.

After a struggle we reached the higher ground just west of the summit of Kosciusko and from there we could see the road from Jindabyne, in New South Wales, up through Charlotte Pass to the summit. At Charlotte Pass there was a large tourist hotel, but between the pass and the summit was just a winding, rough, car track. A couple of kilometres down this track from the summit was the Seaman Hut. Laurie Seaman, an American, and Evan Hayes perished in a blizzard on Kosciusko in 1928. With funds provided by Seaman's parents and the NSW government, a memorial refuge hut was erected in 1929. It was a very well built, stone, two-room hut and we intended to say in it for a week.

Heading directly for the hut across snowgrass, we came to a snowdrift on the side of a slope up which we had to travel. The snowdrift was only about 10 or 12 metres wide. Absurd though it may seem, we were so exhausted we could not climb across it with what we were carrying. We dumped our heavier gear and our skis and stocks, then stamped up over this hard consolidated old snow. From the top of the slope it was downhill to the hut. There we rested for a few hours, then got up, cooked a hurried meal and went back to bed. The next morning we recovered our gear from the snowdrift.

We had a wonderful week in the Seaman Hut. In those days it had beds, mattresses, blankets, a stove and firewood. We walked many miles exploring the region. There was a large snowdrift below the hut on which we practised skiing and, later in the week, there was a heavy snowfall. Luckily, it had not occurred during our climb. At the end of the week we packed up, loaded our gear on our backs, walked uphill to Mount Kosciusko, down into Wilkinsons Valley, up onto the Abbot Range, down Hannels Spur and into the timber. We reached the bottom, undressed and waded through the river, dressed and walked three kilometres downstream, straddled our way across the two-wire bridge and spent the night in the Geehi Hut.

Next morning Geoff took a bridle and got the horse from the paddock. He tethered it outside the hut, but when he tried to place the pack-saddle on its back the horse reared, broke the bridle and raced off into the bush. We never saw it again.

So we jettisoned our food and utensils, rolled our blankets over our shoulders and set out to walk back about 20 kilometres to Waterfall Farm, wading through the icy river eight times on the way. Scammel wasn't worried abut the horse; he said it would come back. So we climbed into the Ford and drove back to Melbourne.

This trip taught me the limit of my physical resources and the sensation of utter exhaustion. And it was the experience gained on Hannels Spur that led me to propose, in 1938, a winter venture on the same route from which I was very lucky to return.

6

ON BOARD HMAS *WYATT EARP*

In 1947 the Australian government established the Australian National Antarctic Research Expeditions (ANARE). In December of that year HMAS Wyatt Earp *was despatched to carry out a reconnaissance of the coast of George V Land, seeking a suitable site for a permanent station on the Antarctic Continent*

After 19 years of voyages through some of the stormiest seas in the world I still consider myself a landlubber. I know something about navigation and a lot about handling a ship in ice, but I can tie only three or four kinds of knots, and I still tend to talk abut windows and floors and walls on ships. Although I never actually become seasick any more, I never quite get my sea legs, and I think my liver is permanently classified as third class as a result of what it has had to put up with. In fact, I am more a ship-hater than a ship-lover.

My voyage on the *Wyatt Earp* was my first experience. My job on the ship was to carry out cosmic ray measurements to investigate variations of the radiation with latitude between Australia and Antarctica. To house the equipment I had arranged for a laboratory to be built to my design on the after end of the boat deck

The expedition was a failure in most respects. The tiny wooden ship (400 tons gross) was quite unsuitable for the task she had been set. Built in Norway in 1919 for North Sea herring fishing, she was bought in 1933 by the US Antarctic explorer, Lincoln Ellsworth. Having made four summer voyages to the Antarctic under his leadership, she was ultimately purchased by the Australian government and, after service with the Royal Australian Navy on the Australian coast during World War II, came to rest in the Torrens River, Port Adelaide, as a hulk.

Early in 1947 the ship was extensively rebuilt, fitted with a new powerful, diesel engine and commissioned with the Royal Australian Navy for Antarctic service with the ANARE.

The bridge superstructure had been moved forward to allow more cabin accommodation amidships. The crew slept in the forecastle. The *Wyatt Earp* could do eight knots and was fitted with sails which, in a stiff following wind, boosted her performance to nine knots!

We sailed from Port Adelaide on a journey which was to confirm for me the worst sailors' superstitions. In a spirit of bravado, I guess, we set out on 13 December and things went wrong right from the start.

On the 14th the main engine failed and the gyro compass ceased to function. On the 15th, the steering chain parted. On the 16th, during lifeboat practice, a davit guy wire broke and four sailors narrowly escaped being flattened when a heavy roll sent the 26-foot whaler crashing against the bulkhead against which they were standing. They barely managed to throw themselves prostrate on the deck in time.

We butted our way to Melbourne against almost continuous storm and I was miserably sick all the way, but had to keep on my feet to maintain my cosmic ray equipment. In Melbourne, a number of repairs were effected and, this time, just to prove we were not superstitious, we sailed on a Friday, 19 December, for Antarctica. Halfway down the bay the engine starting handle jammed and we took three hours to repair it.

On the Saturday we ran into a Force 10 easterly and for 24 hours took a terrific bashing—if we hadn't been able to shelter in the lee of Flinders Island we might have been in serious trouble, because the ship was a shambles.

She rolled 45° to 50° each side of vertical with a roll period of five seconds! All loose objects—and most fastened ones—were cast loose and hurled from side to side. The noise was a series of deafening crashes which continued unabated. Crockery, ship's fittings, personal belongings, books, suitcases, drawers which had burst loose and banging doors which had sprung open, smashed back and forth in cabins and passageways.

To add to the confusion, the ship's timbers sprung leaks just above the waterline, level with the cupboards beneath the bunks, and water seeped in each time the ship rolled. Soon, most cabins were ankle-deep in water and in this floated shoes, clothes, papers, dissolving books, soap and personal gear.

The radio room was indescribable. As well as the usual papers and rubbish, a five-gallon demijohn of distilled water was trying on each roll to torpedo a hole through the bulkhead, while the main radio receiver, which had come adrift, slid half-submerged from side to side of the cabin.

The cook's galley was a treat! At the height of the storm, pots of stew and potatoes disgorged their contents onto the deck and covered it with a cold, greasy, glutinous emulsion in which the cook and his assistant strove vainly for foothold and into which crashed plates, cups, cutlery and the day's supply of tinned provisions.

Lunchtime found us sitting like sand-boys on the wardroom table, bare-footed, with trousers rolled above the knees, sipping soup out of warmed tins while filthy seawater sloshed around under us and the whole place rocked as in a crazy dream!

We hove-to in the shelter of Flinders Island and restored the ship to a manageable condition after 12 hours' work. By then the seas had moderated and we were able to proceed to Hobart, which we reached on Monday, 22 December. Shipwrights came on board to caulk the ship, and engineers to repair various mechanical faults. The gyro was not working.

On Thursday the 25th we set out again, intending to swing compasses in the Derwent, then return the experts to the pier before finally departing for Antarctica. However, when returning to the pier, the engines refused to go astern when required and the ship ploughed her stem through a few metres of pier before coming to rest.

We finally got away—quite undaunted by old wives' tales—on a Friday again—Friday 26th. An hour later we stopped to replace a faulty engine valve, and eight hours later we ran into a Force 10 south-wester! On Monday 29th the steering chain parted again and the gyro gave up for the third time.

This set the pattern for the trip. We found that the water continued to pour into the cabins through cracks just above the waterline every time we rolled.

I lived with two others in a 'black hole of Calcutta' in the centre of the ship, with no portholes and no ventilation except, just above my pillow, an iron grating which opened onto the boat deck above. As the deck was always wet, water dripped through onto my face until I nailed a piece of tin over the opening. Thereafter we had the choice of suffocating overnight with the door shut, or opening it and becoming asphyxiated by the oily diesel fumes—like those from the old Bourke

Street buses—which wafted in from the open engine-room casing next door!

However, we had the only dry cabin—that was until the second officer, in a wet cabin just astern of us, used a brace and bit to bore drainage holes from his cabin at floor level into ours.

We had travelled south about 1600 kilometres when the chief engineer told the captain that the new engines were 'settling down' on their Oregon beds, and that the propeller shaft had a bend in it and was likely to snap at any minute. Because of the side pressure due to this bend, the gland where the shaft passed through the stern timbers to reach the propeller ran hot. To relieve this condition the greasers loosened the gland, whereupon the sea poured in.

The large flywheel in the engine room had only about eight centimetres clearance from the bilges; when the water came in, the flywheel sprayed bilgewater around the engine room like sparks from a catherine-wheel! So they had to tighten up the stern gland and take the chance of the shaft snapping.

When all this was reported back to Navy Office, we received a curt signal, 'Return at once'. So on 1 January 1948 we sadly turned about and headed for Melbourne. Strangely enough, all on board were furious at the time and couldn't understand why such caution should have been exercised back at Melbourne.

After due repairs in Melbourne we finally sailed, but had sense enough to choose a Sunday (8 February) for our departure, and this time succeeded in reaching the pack ice off King George V Land, although we didn't manage to push through to the coast. We did make a running survey of the Balleny Islands, north of Victoria Land, before proceeding to Macquarie Island.

On the morning of Wednesday 24 March we were sailing from Macquarie Island to Melbourne on our return voyage from Antarctica. The weather was atrocious—a Force 9 north-westerly storm into which we headed almost directly. The seas were the renowned huge swells of the Southern Ocean and were as high as any we had seen previously. The *Wyatt Earp* would climb to the towering crest of one, then plunge with sickening velocity—faster than the fastest lift—to the trough, where she would bury her bows in the next swell before shaking free of the engulfing water and laboriously climbing the steep wall ahead. The accelerations on the downward path would be greater than 'g' (the acceleration due to gravity), for the upthrust of the passing wave under the stern would tip the ship and propel its bows downwards so that one's feet would leave the deck. (In bed at night in such waves the bunk falls away from beneath one's body and one catches up with it with a jolt at the bottom of the dip). The ship made a maximum speed of only 1.5 knots throughout that day.

I decided to visit the cosray laboratory to make sure everything was secure with the ship moving so violently. I was wearing my rubber knee-length sea boots because the decks were awash. I climbed the companionway stairs leading from the port side of the after end of the main deck up to the boat deck, then waited for a chance to cross beneath the boom of the trysail to reach the door of my laboratory on the starboard side. From this boom hung a number of short lengths of rope which, when the sail was furled, would be used to lash the canvas to the boom.

I had developed a technique for executing this manoeuvre in bad weather. I would wait until the roll of the ship gave me an uphill run, when I would dash forward to grab one of the short ropes hanging from the boom, then wait while the ship heeled over to the other side. When it rolled again to provide me with another uphill run, I would complete my dash to the door of the laboratory.

This time I reached the boom and was clutching the hanging rope during the downhill roll to starboard. Then a larger wave than usual hurled the ship back to port, to the uphill position. My rubber boots slipped on the wet deck and the violent 'throw' of the ship stretched me out horizontally, feet in the air, as I clung to the rope. Then it was torn from my grasp and I was catapulted longitudinally towards the port railings of the deck.

I was saved by the guardrail around the top of the companionway leading from the deck below. If it had not been for this I would have gone feet-first through the port-side railings into the water, the ship would have continued on and no-one would have known.

As it was, I struck the guard rail with great force and was lucky not to break my femur. I was in much pain, but tremendously relieved to find that I was still on board the ship.

On another occasion, towards the end of our time in Antarctic waters, the ship became heavily iced up and in danger of becoming unstable. We were all asked to go on deck and chip off the ice. However, the wires and ropes in the rigging were also iced up, so that they became cylinders of ice, anything from two and a half to five centimetres in diameter.

Because it was so cold, I was wearing my ANARE cap, a cloth cap I had designed for the expedition men generally. It had ear flaps, a large peak and a padded flat top to help keep one's skull warm. A piece of ice like a belaying pin fell 10 metres from the rigging. It was about half a metre long and five centimetres in diameter, solid ice, and it struck me in the centre of the skull, knocking me to the deck. I was not unconscious, but it was a very heavy blow. If I had not been wearing my padded ANARE cap I would undoubtedly have suffered a fractured skull.

About a year later, the *Wyatt Earp* was sold to a private firm, and Captain Jerry Haines who sailed her under her old name *Wongala* told me of a number of other defects he discovered in the ship—only minor things, like worms in the rudder post—but enough to have spoiled my peace of mind on my first Antarctic journey, had I known!

Postscript

In 1948 I had made two voyages—one to Antarctica and one to Japan—to make cosmic ray observations in order to study the variations in intensity of the rays with latitude.

Towards the end of 1948 I was working at Melbourne University on the collected data, preparing a scientific paper for publication. The data sheets were stored in a simple safe in a wooden hut situated between the Physics School and the University Union House. They comprised numerous pages upon which had been listed the hourly cosmic ray counts from three Geiger-counter arrays in our apparatus, together with certain meteorological data. There were many thousands of entries.

Arriving at the Physics School one morning I was appalled to find that a fire overnight had destroyed the hut. Wandering disconsolately through the ashes and rubble, I reflected that the results of two years' work had been destroyed. The door of the safe had sprung open and piles of charred paper were protruding.

As anyone knows who has tried to burn old newspapers in an incinerator, piles of papers do not burn easily. Picking up a few sheets from the charred heap I was surprised to find that, apart from those in the top few layers, the other sheets, although charred around the edges and saturated by water from the fire hoses, were undamaged.

My project colleague Colin Mackenzie and I separated all the sheets and spread them out over the floors of a couple of rooms in the Physics School to dry. Next day we used scissors to cut around the sheets to remove the charred edges. Because of wide margins on the paper, the columns of figures on each page were unscarred.

In 1948 ballpoint pens had just reached Australia and I had taken some, in preference to fountain pens, purely for convenience. It was fortunate that I had, for ballpoints use indelible ink, whereas the ink in a fountain pen is water soluble. The ballpoint pen entries on the data sheets were unaffected by the fire hose water.

All in all, we salvaged about 95 per cent of our data, and as a result of our good luck, we were able to process the data and produce two scientific papers. As far as I know, the cosmic ray papers were the first ever published as a result of a sea voyage to the Antarctic.

Recently, Dr Bob Jacklyn retired from the Physics Department at the University of Tasmania. For approximately 40 years he has continued the cosmic ray work that began in 1948. In May 1992 he advised me that he had found a box containing CR data sheets amongst old records in the Department. They constitute the complete set of data sheets from my two voyages, charred edges and all! I was delighted when he presented them to me.

* * *

Sections of this episode were published as part of Phillip Law's book *The Antarctic Voyage of Wyatt Earp* (Allen & Unwin, 1995).

Iceberg. The tide mark around the foot shows that this one is aground. *(P. G. Law)*

Geoff and Phil Law at the Seaman Hut, Mount Kosciusko. *(A. Rigby)*

The Geehi Hut and Phillip Law. *(P.G. Law)*

Left: Geoff Law ascending Hannels Spur, May, 1936.

Right: Phillip Law pushing his way through the foilage up Hannels Spur in August 1938. The track is completely obscured by winter snow.

First ascent of Mount Henderson February 1955. Left to right: Law, Béchervaise, Shaw. Masson Range in background. *(A. Hall)*

Kirkby and McGregor landing on Lewis Island. The terminal ice cliffs of the continent are in the background. *(P.G. Law)*

Phillip Law leads the sledging party towards Lichen Island. *(J. Béchervaise)*

The camp on sea ice at the Gap. *(P. G. Law)*

The camp on Lichen Island *(P. G. Law)*

Near miss—*Magga Dan* slides past the face of an iceberg that the ship narrowly missed. (See Chapter 19) *(P.G. Law)*

Good morning! *(P.G. Law)*

Glacier icebergs. *(P.G. Law)*

DUKW at Mt Henderson *(P. G. Law)*

Weasel party travelling over fast ice towards Larsemann Hills. *(P.G. Law)*

Loading a DUKW at Macquarie Island. *(P.G. Law)*

7

SCUBA DIVING IN THE ANTARCTIC

Over the years 1956 to 1962, I experimented with diving equipment while on Antarctic voyages.

One is led to reflect, after an Antarctic voyage, upon the vulnerability of a ship in a polar environment. There is, first, the matter of isolation; if any misfortune eventuates, one is a long way from help of any kind. Before the International Geophysical Year (IGY), ANARE ships were the only ones in the vast expanse of ocean fringing the Antarctic continent outside the region of the Antarctic Peninsula (south of South America). After the IGY it became possible to call for aid from American or Russian icebreakers, but they were generally several thousand kilometres distant and, even if they could help, they would take days to come.

Second, any ship is merely a 'tin can' when viewed in relation to the immense forces involved when pack ice is under pressure from winds and currents, or when the ship collides with bergy bits hurled against the hull by great waves or, worse still, if the ship is forced against an iceberg by hurricane winds. A ship close to an iceberg or the edge of an ice shelf is always at risk from the collapse of part of the ice face. On one occasion, flying over the 20-metre terminal cliffs of an ice shelf, I saw a section disintegrate: several hundred metres of the ice front 'exploded', hurling thousands of tonnes of ice into the air and hundreds of metres out into the ocean, while the disturbance created great waves and turbulence in the vicinity. Any ship close to such an event would have little chance.

But there are other less dramatic risks. It occurred to me that it would not need much to happen to the submerged portion of a ship's hull to produce a serious emergency. For example, the simple matter of a steel hawser becoming entangled with the ship's propeller might immobilise the vessel because the low temperature of the sea water would prevent anyone from attempting repairs from outside the hull.

I decided it would be worthwhile to obtain diving gear and to prepare myself to give assistance in any such emergency. An adequately equipped diver, protected against the cold could, for example, wield an oxyacetylene torch underwater to good effect in certain circumstances.

When I first took up this matter, in 1956, the only diving suit available to me was of the type used by Navy and commercial divers—a closed, heavy, canvas suit and a bronze head cowl with a screw-down metal-bordered 'window' to look through. An air line supplied compressed oxygen from a container 'bottle' hanging on one's chest. I borrowed this equipment from the RAN and tried it during one voyage, in 1956, but found it quite unsuitable. It was heavy and clumsy. I needed assistance to don the gear and to walk on deck, and the time taken was unacceptably long.

I shall not forget my first dive, which took place in Mawson Harbour on 1 March 1956. It was probably the first dive with underwater equipment ever recorded in Antarctica. After dinner that night John Bunt, Mawson biologist, helped me lug the gear from the hold up to the saloon where, with a certain amount of trepidation, I climbed into it after first donning heavy woollen underwear.

To help keep my head warm, I had decided to wear a woollen balaclava under the head gear, but in pulling the latter over my head the balaclava became wound around my face and I could not breathe. Bunt had to drag off my head cowl and I cast aside the balaclava.

After replacing the cowl, Bunt screwed down the brass-bound glass visor in front of my face. It seemed almost too small to allow air in, even if it were open, but with it closed and my having to breathe oxygen through a tube, the claustrophobic effect was almost unbearable. It needed all my self-control not to scream 'Get me out of here!' My heart beat furiously and my breathing rate increased to the point of panting. I told myself to calm down and relax, and walked around the saloon for a while to accustom myself to the breathing.

When I had recovered, my assistants strapped on the accessories—the compressed oxygen bottle in front, with its control valve, the lead belt, and the rubber bag, or reservoir, which acted as an 'accumulator' or 'buffer' between the oxygen bottle and my mouthpiece. This bag would not fit easily over my head and when it was dragged on it nearly pushed in my face.

Finally I was fully rigged, but then became too hot in the warmed saloon, so went out on deck, where I squatted down on a bollard, relaxed and practised breathing for about 10 minutes.

I looked over the side of the ship at the rope ladder dangling four metres down and decided that there must be a better way for a heavily laden person to enter and leave the water; so I asked the Army crew to launch a DUKW. I finally clambered down the ladder into the DUKW, belayed from the deck by a rope around my waist.

The DUKW drove over to the shore and I scrambled out into about two metres of water with a safety rope around my waist. When I tried to dive beneath the surface I could not—I just floated on top, with my suit full of air. So I asked the DUKW driver to move out into the harbour while I floated face-down, holding onto the DUKW with one hand.

I was fascinated. The water was thick with myriads of small white jellyfish, a few centimetres in diameter, some of which were luminescent. There were a few large purple jellyfish and the bottom of the harbour was smooth, bare rock.

My body was quite warm and I was breathing easily, but my hands were giving trouble. My left hand wore a canvas mitten, but it was inadequate as insulation in the freezing water. My right hand had no mitten, for I was using it to hang onto the DUKW, and it soon froze up.

I signalled to be helped into the DUKW and ready hands hauled in my safety line as I tried to pull myself in over the gunwale. Suddenly I could not breathe. I waved an arm, pointed to my glass visor, and yelled 'I can't breathe'. I thought they'd never hear me through the thick glass and wouldn't know what was wrong. Imprisoned in this canvas and metal suit, unable to breathe and unable to communicate, it was a terrifying moment. It seemed an age before anything happened, but it could not have been more than a few seconds.

Strong arms lifted me into the DUKW and Bunt unscrewed a valve in front of my nose that let in fresh air. At the same time the oxygen flow recommenced.

The reason for the trouble was simple. As I rose out of the water my weight increased and so did the tension on my safety line. This line just happened to cross over the tube leading from my chest bottle of oxygen to my face and, as I leaned over the gunwale, the tension on the line and my weight, both pressing on the oxygen tube, blocked the flow. Once over the gunwale and into the DUKW the pressures were eased.

All this sounds very primitive today, and so it was. It was several years before I learned of Jacques Cousteau's invention of the aqualung for scuba diving. In the meantime I took the RAN equipment on each voyage in case an emergency arose that required it. Fortunately none did.

The question of adequate hand covering, though, was to concern me for quite a while. Even with my first use of an aqualung, I still had not solved the glove difficulty. I had used various combinations of woollen mittens, canvas over-gloves, rubber gloves, etc. but could not get the right mix of flexibility and insulation. It was some time after wetsuits became available before mittens of the same sponge-rubber material were manufactured, no doubt because there was little demand for gloves for use in freezing water.

Cousteau's invention of the aqualung opened up new prospects for underwater activity in the Antarctic. The wetsuit seemed to me to offer possibilities for protection in ice-cold water. It would allow ingress of water but, once the water was trapped inside the suit, I predicted that it would rapidly warm to body temperature and, thereafter, one would not be uncomfortably cold.

So I enrolled in a diving class at the Melbourne City Baths and was taught the use of the equipment, its maintenance and safety procedures. The Antarctic Division purchased the necessary gear and a compressor to refill exhausted air cylinders. I then had some practice in Port Phillip Bay, but because of the mild water temperatures I did not bother to wear the head covering

Early in 1962 we were relieving the Wilkes Station in the ship *Thala Dan*. On 11 January, our second day at Wilkes, I decided to try out my scuba gear. I donned a woollen singlet and long woollen underpants, then dragged on the wetsuit. The suit of sponge rubber was skin-tight over the underclothes. The cowl of sponge rubber was even tighter, restricting the veins around my head and neck and making breathing difficult. It was stupid of me not to have tried it out in Melbourne. I pulled on sponge rubber mittens and went for a surface swim from the deck of a DUKW. I found I could keep warm. Trickles of cold water entered via a zip, but soon heated up

Clambering back into the DUKW, I hauled onto my back the compressed air cylinder and adjusted the face mask, with window and mouthpiece, over my cowl. The result was highly claustrophobic, mainly because of the cowl restriction, so I took off the face mask, removed the sponge rubber cowl, and donned the mask again. That was much more comfortable, so I entered the water and dived. The breathing was now all right but after 30 seconds my bald head and ears ached so painfully from the cold that I had to stop the dive.

I had no option but to wear the rubber cowl. When I had again fitted it and pulled on the face mask, I sat and tried to relax and conquer my claustrophobia. I resisted the urge to pant and gasp for air and forced myself to breathe slowly and regularly. After several minutes I had enough confidence to enter the water and dive down under the DUKW. The effort of swimming kept me warm.

Once under the surface the usual fascination of the underwater world caused me to forget the discomfort and I swam around confidently beneath the keel of the ship. The sea was opaque with plankton, like a snowstorm under water or as though tissue paper had been roughly torn into pieces as large as a fingernail and allowed to float in millions beneath the surface.

The safety rope around my waist was a damned nuisance. Also the air bottle on my back slipped down towards my head when I was upside down and, even when upright, I could not put my head right back to look up because I banged the back of my head on the valve on the top of the bottle.

After a bit more experimenting, I gave up and was taken back to the ship. The coldest part of the whole exercise had been in the DUKW and on the ship's deck after leaving the water—the wind was blowing and I was no longer exerting myself.

At the end of all this I was very tired—more exhausted nervously than physically. Using equipment with which I was relatively unfamiliar in unusual circumstances and forcing myself to do certain things against all my natural reactions apparently took toll of my nerves without my noticing until later, when I relaxed.

Despite all this I was pleased with the test. I now felt confident enough to use the equipment in an emergency.

Thirteen days later, after leaving Chick Island, where we had established an automatic weather station. *Thala Dan* tied up alongside a large ice floe just off the iceberg tongue jutting out from Cape Mikhaylov. While the helicopters were away taking a survey party to the cape, I decided to try the scuba gear again.

Before doing so I approached the captain and said, 'Have you seen any killer whales around? In my black wetsuit I look like a seal, like enough for a whale not to notice the difference.' He said, 'If I were scuba diving here I would have a lot more serious worries than the chance of a killer whale. In any case, we have just stopped the ship. I guess the sound of the engines will have frightened away any in the vicinity.'

This time I did not wear the heavy woollen underclothing, just a pair of briefs and a cotton singlet. Over these the wetsuit was not nearly as tight. I strapped on the air bottle, climbed down the ladder from the ship to the floe, put on my mask and flippers, pulled on my sponge-rubber mittens, then negotiated some steps that a sailor had cut at the edge of the ice and entered the water.

Although I became quite wet I did not feel cold. On my first dive my ears ached with the pressure, so I returned to near the surface and swam around for a while to take things quietly and relax. My next attempt was successful so I proceeded beneath the ice floe.

The water was crystal clear; however, the sun was low on the horizon on the other side of the floe, so beneath the floe it was too dark to see anything much. Some water entered my mask and, with my hands encased in mittens, I could not lift the edge of the mask to blow it out, so I returned to the surface for the job.

Before entering the water I had told my helper, Tom Harwood, that if I gave a couple of hard tugs on the safety rope he was to haul me in, fast!

I swam around the ship and under the edge of the floe for about 10 minutes and then began to have trouble breathing. Thinking the compressed-air bottle might not be fully filled, I switched to the emergency supply, but without improvement. Suddenly, at a depth of about 10 metres beneath the propeller of the ship, the air cut off completely and I couldn't breathe. I shot up to the surface and tore off my mask, taking great gulps of air. I then gave two tugs of the rope, but Tom did not feel them—my sudden rise had created slack rope that he was busy winding in. So I swam hard with my unmasked face above the water and then, to my dismay, saw a killer whale surface near the ship only 30 yards away. I again tugged on the rope and this time Tom felt it. He looked at me and I gestured towards the whale, whereupon he hauled hard and I planed back along the surface to the steps like a surfer who has caught a wave. The whale dived and we did not see it again.

After I had returned to my cabin I pulled apart the demand valve of the aqualung and found crystals of ice. It became clear why the air supply had cut out. My Australian-made aqualung had the demand valve mounted in front of the mouth. Expired air, moist from the lungs, was expelled through this valve. The metal valve was in contact with sea water at a temperature, amongst the ice floes, of around 29-30°F; thus the water vapour from my breath condensed, froze and clogged the valve.

By contrast, the Cousteau aqualung has the valve mounted behind one's head. Two rubber tubes lead from the mouth, one either side of the face, around to the valve at the back. I think this was done so that bubbles from expired breath would not rise in front of the diver's face and obstruct his view. The tubes have airways more than a centimetre in diameter. Any expired moisture would freeze in these tubes before the air reaches the demand valve. It would take a long time for the ice to build up sufficiently to block the thick tubes

8

YOUTHFUL MEMORIES

Mount Buller Ski Fall

I was in the habit of proceeding to Mount Buller, about 230 kilometres north of Melbourne, for odd weekends during winter for skiing. One weekend in 1958 I was staying in the lodge of the Alpine Club of Victoria. It was a beautiful morning, calm and sunny. Arising before the others were awake, I donned my skis and pushed out across the sparkling snow.

After a couple of downhill runs I decided to move further south to the next gully. This involved traversing a fairly steep slope through snow-gum trees, over snow which, at that early hour, was still hard-frozen. My ski edges were cutting nicely into this hard snow as I gently slid along, slightly downhill, underneath various trees.

Suddenly, under one snow-gum, my feet went from under me. Overnight, dew had formed on the boughs of the tree and drops of water had fallen onto the snow and frozen as spheres so that, on the firm snow beneath the snow-gum, there was a sprinkling of marbles of ice that acted like ball-bearings under my skis. The skis simply slid sideways; I fell flat on the ice and began to glissade downhill. The icy slope became steadily steeper and I found myself sliding faster and quite out of control. I held a ski-stock in both hands and dug its point into the snow to try to slow down, but all this managed to do was to swing me around so I was sliding headfirst on my back. As I saw branches of snow-gums whiz past, I kept throwing my hands up hoping to grasp something. Finally, by sheer good luck, I clutched a bough and managed to hang on although my arms were nearly wrenched from their sockets. I gathered my skis in the right position, managed to stand up, then looked back up the slope.

On the snow were the tracks of my long slide. They ran straight down the slope, past and through numerous snow-gums and jagged outcrops of rock. Beneath me, a mere 20 metres further on, was a minor cliff which fell to further rocks below.

The only straight line through the various obstacles above me on the snow slope was the line of my slide.

Nigretta Falls

In 1925 I was attending high school at Hamilton, in the Western District of Victoria.

Some 20-odd kilometres out of Hamilton, the Wannon River flows over the Nigretta Falls. These falls are a minor tourist attraction, providing quite an attractive hike for sightseers who can walk in a roundabout way from the high land on either side down to the foot of the gorge and then back upstream to where the thunder of the water coming over the falls is in their ears and the mist from the falls in their faces.

At the age of 13, I and four others—Ron Stewart, Bruce Robson, 'Ginger'(Bob) Eddy and my brother Geoff went out to Nigretta Falls on our bicycles. The whole area was still pretty much as it had always been, in a wild state, with a few poorly marked tracks where feet had worn paths. Parking our bicycles, we wandered down into the gorge and then about half a mile upstream to reach the foot of the falls. When we were ready to leave, we decided it would be a shorter track to our bicycles if we could somehow just scale the cliff to one side of the falls. Accordingly we scrambled up a steep slope to where there was a sort of 'chimney'—a cleft in the cliff above us that was fairly steep and choked with boulders. The five of us successively entered this cleft, which was very near the top of the cliff. Ron was just ahead of me, Bruce was behind. Geoff and Ginger were leading. Higher up a boulder about a metre in diameter choked the chimney. The two leaders were able to squeeze through past the edge of it. As Ron pulled his way past it, I saw the boulder move, so I flattened myself against the side of the chimney. It rolled past me and struck Bruce, knocking him from his foothold. He followed the boulder down the cliff in a series of bouncing somersaults, through a rain of smaller boulders and rocks set free when the large boulder was dislodged.

I still remember my horror at seeing his body bouncing down what was a 60° slope at the foot of the cliff we were climbing and finishing huddled in a heap at the bottom. My God, I thought, he must be dead Then, to my amazement, I saw this little figure hunch up on its hands and knees and crawl slowly to one side, with boulders and rocks still whizzing past from above.

Bruce's leg was broken just above the knee, and he had a deep gash in the thigh. We carried him downstream, mainly though the water because it was easier, and then up to the top. There we met some other visitors to the falls who were able to take him in their car to the Hamilton Hospital. Our main fear was fronting up to his parents, for we felt guilty about having led him foolishly upon such a venture.

Mitta Mitta River

I was born in Tallangatta, a small town in northeastern Victoria. The following incident occurred in about 1920.

The Mitta Mitta River today flows into the Hume Reservoir near Tallangatta. However, when I was a young boy holidaying in Tallangatta, the river flowed on to join the Murray River at Albury. At Tallangatta the Mitta was fast and deep and winding, with numerous sandbars, and its banks were clothed with glorious willow trees whose fronds dipped into the water that swirled beneath. The local swimmers had erected a diving board from a 12-foot bank over their swimming pool, but a late September flood had washed out the board and that summer it was lying on the bank.

One of our favourite pastimes was to swim to the fast-flowing centre of the river and drift downstream. This was a dangerous practice because of snags in the river which could easily tear your body if you were washed over them.

On one occasion a friend and I decided that the safest thing to do would be to lie on the diving board, which was about three metres long and seven centimetres thick. So we dragged the board down into the river, lay upon it and paddled with our arms. We were carried rapidly downstream, swirling in under the fronds of the willows and out into midstream again, with the speed thrillingly obvious because our noses were only a few inches above the water. After a couple of miles we grounded the board, left it there, and walked back gingerly through the thistles and weeds, then over the sandbanks, to where we had started.

However, prior to this experience I had risked the trip without the board. Floating downstream on my back, I was enjoying the effortless motion when a snag caught in the back of my old-fashioned, one-piece woollen swimming togs. I was suddenly stopped in midstream and the river surged over me. My view of the sky was suddenly obliterated by a swirling turbulence of green and white and I coughed and choked as water invaded my mouth and nostrils. I twisted and turned as I desperately attempted to grab the branch beneath me, but could not reach it. Panic-stricken, with my breath almost exhausted, I made one last frantic convulsion, my togs tore and I drifted free, gulping for air.

It is a perfect example of how some unexpected and accidental incident over which one has no personal control can, under certain circumstances, be fatal.

9

BY MOTOR LAUNCH TO AMUNDSEN BAY

Enderby Land marks the western limit of Australian Antarctic Territory. At Proclamation Island just off its northern extremity, Sir Douglas Mawson had in 1930 made one of the three new landings accomplished by the British Australian New Zealand Antarctic Expedition (BANZARE). Shortly before, Captain Hjalmer Riiser-Larsen had led a Norwegian expedition to land on an island further west (off Cape Ann) and the expedition aircraft had made a reconnaissance of the coast. But it was left to the ANARE in 1958 to make the first landings on the continent and to explore and map the detail in this region.

In February 1958 we approached Enderby Land from the east in the ship *Thala Dan* (Captain Kaj Hindberg) and sailed along the coast past Proclamation Island and Cape Ann. I hoped that, with luck, *Thala* could push into Amundsen Bay, where no ship had ever been. My voyage diary has the following entry:

Friday 14 February 1958
Noon—66°31S, 5°11E

Up at 0320 hours—weather only fair—cloudy, NE wind (light 8 knots) but a NW swell and the barometer dropping. I decided to try and hurry through a couple of aircraft flights, so arranged to hoist the plane outboard at 0500; but it was 0600 when it went out, owing to delays in rigging the 'jumbo' derrick satisfactorily. In the swell the unloading operation was a bit ticklish and I didn't like it much.

The plane flew for about an hour while I tried to persuade the captain to tuck the ship close under the lee of an iceberg to avoid the swell. No luck—he wouldn't go within ¼ mile of it.

When the plane returned, the wind and swell were bad and we had trouble getting the plane alongside. I decided to lift it on board. This we did with some risk and trouble. I was glad to have it safely hoisted. Any future flight was out of the question in the sea and wind.

The boys told me they had found some new islets close to the continental shore just north of the entrance to Amundsen Bay and that a narrow lead of water along the ice cliffs would allow the ship to reach them. The captain refused to go, saying he would be 'blocked in' (what rot!) so I said 'Take me to the entrance and I'll go the rest in the launch.'

So now at 0900 we were proceeding to do that. Adams was preparing our material for survival and the scientists were getting ready. I decided to take Osmar White (journalist) too.

Soon after, the captain came to me and said that it was quite open, pointed out the islets nestling under the continental ice cliffs, and asked me did I wish to go to the islets or to go as far in as he could get. Here I made a mistake—I said 'I'll settle for the islets. If I want to go further I can use the launch.'

My main reason for doing this was that time was going on and I wished to get the surveyor, Knuckey, ashore before noon for his astrofix. I should have asked to go as far as possible, as events turned out.

As soon as we stopped I had the launch put over, containing White, Cook, Jesson, Knuckey, McLeod, Lucas, Stinear and myself, and by this time it was 1115 hours. We cruised in towards a group of islets—low, rocky, elongated, black—in fairly rough sea whipped up by a 20-knot NE wind. One islet was covered with Adélie penguins but I sailed around it without finding any place to land in the sea that was running.

We then went across to the most northerly islet and found similar difficulty. Noon came and we still were not ashore. I was becoming a bit desperate and, when I found an ice ledge about six feet above water level, under which the water was fairly deep, I took my courage by the throat and took the boat in.

The next ten minutes were rather hectic. The launch heaved up and down in a four-foot surge and bashed against the ice. Sometimes it would sink in a trough and slide its stem in under the ice ledge, then rise and grind into the ice, breaking off great chunks to fall on the deck.

I stayed at the wheel, alternately backing and forwarding to try to maintain the launch's position. From there I had to relay numerous messages through Os White to the men unloading their gear, for the scientists had no 'nous' and were doing none of the right things that an experienced boat crew would have done. Eventually an anchor was put into a snow drift ashore and I decided to put out only Knuckey with his instruments and Stinear to act as his assistant. I set a man holding rope buffers to absorb most of the shock but the launch was taking a beating and as soon as I could we pulled out. I went back to the ship and put the scientists on board, then took on Lucas, White, Donovan and Adams as my new crew.

Knuckey was set up on the crest of the islet by 1230 hours and obtained a good fix. He built a cairn and deposited a cylinder I had given him.

I asked Thompson to take the pontoon to the island and land six drums of avgas at a depot, also to arrange further flights for the RAAF if the seas moderated. We set out in the launch at 1245 hours to go to Amundsen Bay. The weather looked a bit brighter but there was still a strong NE wind and a lot of cloud. I set the echo sounder running and took records of the depths as we ran along the coast about a kilometre off the ice cliffs of the continent, past the group of islands upon one of which Knuckey could be seen working, and past several skerries further offshore. As we went along conditions improved. Ahead could be seen the point where the RAAF had said the open water was almost closed by icebergs, and the pack ice edge and the ice cliffs converged. Also ahead lay the interesting skyline of mountains fringing the eastern and southern sides of the bay. The depth was uniformly 65-70 fathoms. To starboard was the ice dome of 'White Island' (as I think it is marked by the Norwegians).

In fact there were two iceberg gaps ahead, the first one quite easily negotiated and the second half-blocked by floes and brash. When negotiating the second gap, where the ice cliffs turned to run ENE, I noted that the NE wind was holding the heavy ice away from the cliff and thought that, if the wind were to swing south, the gap would close up. From then until our return I was worried by this possibility. We pushed on with the sturdy launch butting through brash and light pack ice floes.

Around this corner the view was marvellous. About a kilometre ahead we could see the open pack give way to clear water. When we reached this water we found ourselves in a great expanse of blue bay, smooth as a lake, fringed by mountains of alpine grandeur on which the sun shone brightly. The scene was the best alpine one I had seen in Antarctica, and I took numerous photographs.

We decided to go to the large peak which was the most westerly of those along the northern coast of the bay, although it looked at least five miles away. We were exhilarated by the scenery. It was as though we had sailed our launch around a corner and suddenly emerged into a Swiss lake!

Although the weather where we were was glorious and I thought the depression must have passed, I noticed a little later that a 'front' of white, heavy clouds, backed by dark ominous sky, was moving in from the west.

Until now we had held radio schedules each hour with the ship but, once around the corner of the continent, our signals could not get through at all and theirs were unreadable. I pondered whether conditions at the ship might have deteriorated badly but, although time was pressing, I was determined to reach the mountain ashore and make a landing, so refused to think of possibilities like weather and iceberg gaps.

The large mountain, which I have decided to name Mount Sheila[1] in honour of Thompson's wife (who that day gave him a daughter), was strikingly peaked and similar in general shape to Everest. Its sheer sides were stratified with a dip of about 20° to the east and appeared to contain much metamorphosed sedimentary rock. Dykes of dolerite on the main peak made it look like the cross-section of a classical volcano.

The next peak exhibited striking rock faces, ridges and buttresses—magnificent rock-climbing stuff! We were all sorry we could not spend a week or two here, exploring and climbing. A large glacier wound back into the interior, east of this second peak, and it was flanked on each side by further ranges. There was much rich-coloured brown rock, warmed in hue by the direct sunlight to make glorious Kodachrome material. (Unfortunately, I must have knocked the speed setting on my Leica exposure meter and obtained consistently high readings which might have spoiled some of my shots even though I think I discounted the readings somewhat.)

On our port side the steep ice cliffs gave way to an ice slope, weathered and furrowed remarkably by the passage of thousands of Adélie penguins, which led down to a low ice ledge with deep water right to its margin. We cruised in and I moored against it just as though it were a private jetty. Time—1715 hours. This was 4½ hours after leaving the ship and I estimated we had come about 40 kilometres. I told the men we should try to be back to the boat by 1745.

---

1. No. We later found it had been named Mount Riiser-Larsen, by Douglas Mawson after having been first sighted from the air by the Norwegians.

Hurriedly grabbing our cameras, we walked quickly across a narrow moraine valley and climbed a small headland about 30 metres high to the east of our anchorage. It was composed of the usual granitic gneiss, much finer grained than at Mawson and Davis, with some garnet and cream feldspar and quartz.

Near where we disembarked there were two boulders heavily flaked with stained brown iron ore. The moraine erratics were lacking in variety and seemed to comprise only country rock (very different from the varied assortments to be found at Mawson and Davis).

At the top we stayed about ten minutes. I took hurried photos, including a panorama of Leica shots from west through north to east, and cine shots of men building the cairn. The cairn was only a small one, for rocks were not abundant and time was pressing.

The view was worth millions. Unfortunately the sun disappeared behind a cloud just after we reached the top, so I failed to get the colour telephoto shot I wanted of the stratified layers in the western peak.

Reluctantly we hurried down. Amongst the profuse lichens in a sheltered spot we found a hairy, long-fringed lichen of the sort they feed to reindeers on Kerguelen. It was the first time I had seen it in Antarctica. I brought a rock back with some adhering to it. There were of course the other usual kinds—black, grey, yellow, green, orange, red—as well as green algae and moss. On one side of the ice near the moraine valley there was a number of moulting emperor penguins.

The second mountain had a magnificent scree slope stretching from its summit to sea level. Away inland, near the head of the long glacier, was a remarkable tor. The vertical cliffs must have been more than 1000 feet sheer!

To the east of this, Tula Range faded out. There was a gap of plateau ice before the mountains started once again and ran around the south side of the Bay. The most conspicuous mountain was one running NE-SW, long and low and ice-free and brown, like a dormant rat.

At 1800 we reluctantly chugged away and made full speed back to the gap. While on the headland my fears had been built up by feeling a southerly wind fanning my face and seeing it rippling the water below. However, when we reached the gap, the wind was still NE and the ice had not changed much, although it was closer and we had more trouble penetrating it than we had had on our way out. We would have been in serious trouble an hour later!

From then on all was straight going. We were again in touch with the ship by radio and I gave them 10 p.m. as our ETA. The lane of open water along the ice cliffs was wide and reasonably smooth and we could have flown the plane from it with no trouble. I was sorry I had not taken the captain's offer to go in there.

Soon after passing through the gap the wind freshened and soon rose to a full NE storm at about 50 knots. It was dark and the launch was pitching violently as it butted its way directly into the steep waves, its headlights beaming first at the sky then at the foam-streaked troughs of the swells.

Hanging onto the wheel, legs straddled wide and knees bent, kept me in place, but the others had difficulty avoiding being thrown around in the boat, and were repeatedly drenched with icy spray. I had difficulty avoiding vomiting, for I became seasick. We just hoped our reliable old diesel engine would not fail us, for if it did there would be nothing anyone could do to save us.

At last we saw the ship's lights and the glare of a searchlight that they shone in our direction. It seemed an age before we were alongside and, even then, our troubles were not over. The launch bumped and jarred against the sides of the pitching ship. Sailors dropped hooks for us to attach to the bows and stern so that the launch could be hoisted aboard but with the boat rising and falling six to eight feet this was not easy. One hook became attached astern as the one at the bows failed to fit into place, resulting in the boat being up-ended by the next wave, suspended for a moment almost vertically from the ship's derrick and then being dropped with a big splash down into the next trough. The men trying to attach the hooks were in great danger of being thrown into the sea, particularly those on the decking in front of the cabin where only a small rail 10 cm high acted as protection.

It was a great relief when, after 10 minutes of being thrown violently around, the hooks were in place and the boat began to lift. It struck two wave crests and crashed against the ship's hull, then we were up and swinging free and soon were dropped safely on *Thala's* deck.

It was 2230 hours.

I found on board that Thompson had been unable to find a place to land the drums of petrol, so had not taken the pontoon in. Knuckey had got his fix and Thompson had picked up him and Stinear from the islet with the ship's motor boat without difficulty, as the boat was small and the tide was in. The seas had remained too rough for flying.

Dinner was waiting for me and I for it. I had been up since 0320 hours and had had only one meal—breakfast—apart from some chocolate, an orange, a cup of soup and some biscuits on the launch.

After dining I told the captain he could proceed to Mawson and he started immediately. I collapsed into bed at 0300 hours after an unsuccessful attempt to write up this diary.

I was very pleased with the day's work. We had made two unique landings from the sea, the one at Mount Sheila being unlikely to be repeated for many years. What a treasure the new launch had proved to be!

10

ICE BREAKOUT AT LARSEMANN HILLS

The Larsemann Hills constitute an ice-free area of the continent roughly 100 kilometres south of Davis Station. Up till the time of this account, 1955, no-one had ever landed there from the sea. This chapter is taken straight from my voyage diary.

Off Larsemann Hills—In fast ice at position: 60°10.5S 76°02E
Wind—Practically calm NE 1
Barometer 989.0
Air temperature 28.0°F
Sea temperature ?
Weather—Perfect, cloudless
Sailed from Vestfold Hills 0428 hours.
My lucky day—and what luck! (See later.)

I was called at 0415 hours when the *Kista Dan* set sail from Vestfold Hills for Larsemann Hills. I worked before breakfast writing up my diary for the last two days and preparing for today's programme.

We met fast-ice north of Larsemann Hills at 1015 hours. It was from three to five feet thick, and all hard, no slushy rotten stuff underneath as at Mawson. I visited the crow's-nest and spoke to the captain and we decided we could not break through such ice. Magnificent views all around—across to chocolate-coloured hills and mountains to the east, continent and icebergs to the west, Mount Caroline Mickelsen and icebergs to the south, and all stuck fast with sea ice.

We decided to try a lead which the captain could see to the east. We sailed around to it and I got some fine cine shots of the ship breaking ice on the way. We followed the lead in as far as we could and had to tie up in the fast ice when we reached our limit. This was at 1330 hours. At 0930 I had had a meeting of the men, congratulated them on their work at Vestfold Hills and explained the day's programme. Then I set them packing and sorting and labelling specimens.

I decided to try a weasel dash across the fast ice to the Larsemann Hills about 20 kilometres away. After lunch we had the weasels and caravans hoisted out (of the hold) and put back the ones we didn't require. We chose one weasel, one caravan and a dog sledge. Elliott and Shaw worked hard and got together everything we needed for an 11-man party. I planned to send the weasel straight back to the ship the moment it had landed us and to call it on the radio to come and pick us up the following night. Our team was: Law, Béchervaise, Oldham, Lodwick, Lacey, Krohn, Lowe, Hungerford, Macklin, Gowlett and Power. Gowlett was to drive the weasel and Power to come as companion and assistant for Gowlett on the home journey.

We got away, beautifully organised, at 1730 hours in bright warm sunshine over five centimetres of lovely powder snow lying on the blue fast ice. Just before starting I found that the boys had no chain block, no axe, no ropes (except wire rope), no steel bars to fix in ice. We borrowed a chain block from the engineer, a couple of pulley blocks from the ship, an axe and rope from Thompson, and then we were set.

Soon after leaving we ran through two weak spots in the ice and nearly stranded the weasel in the first 400 metres. Béchervaise and I did a lot of probing but it was difficult to tell the strength of the sea ice. Finally we just rode on the weasel, I directed Gowlett purely from the look of the ice, and when things looked suspicious we jumped down and probed.

About 800 metres from ship we sank dangerously in three lurches but pulled through and left three pools of salt water. We put a marker up there.

When about 3 kilometres from the ship we came across a crack a metre wide full of water. Béchervaise wanted me to use the bridging timbers to cross it, but I felt that it might widen and leave us stranded, so we followed the crack 800 metres to the south and found it narrowed down to a mere crack; but there were half a dozen weak crack areas converging there, with seals and blow holes, and icebergs close by, so I did not like the area much.

I then surveyed our objective through binoculars from the top of the weasel and decided to call off the project. There was blue water near the island which was our objective. All the others, except two small ones about 30 kilometres away, had suspicious cracks (or what looked as though they might be cracks) across between us and them, so I decided to give Larsemann Hills away. (There was some grumbling amongst my colleagues, who thought that, in such fine weather, we should have risked pushing on.)

We got back about 2000 hours and nearly lost the weasel 30 metres from the ship, when it broke through and left a hole two metres long but somehow kept going and got out of it. The captain suggested I try on foot next day but I told him I would not, as there was less chance of crossing watercracks on foot than by weasel. I was being particularly cautious (for me!) about (a) stranding a weasel inaccessibly on the east side of a north-south crack, and (b) having men stranded out of reach of the ship. I think everyone thought I was a bit of a spoilsport to call the whole thing off after all our trouble.

We had dinner about 2040 hours after I had done a lot of photography in the lovely late sunshine. There was a Ross seal on the ice near the ship so I arranged for it to be shot and collected.

Looking out of the porthole of my cabin after dinner I saw cracks in the ice. Then Béchervaise returned from a walk near the iceberg and said cracks were opening up. And did they open! For no apparent reason.

Soon the whole bay of ice had broken up—in less than 20 minutes! The funniest thing was to see disconnected floes floating around separated by great lanes of water, each floe bearing the track marks of our weasel! We were lucky to have had the weasels loaded again on the *Kista Dan* before the break-up. And still not a breath of wind!

I still feel cold at the thought of how narrowly we missed trouble. Three members of the crew arrived back from a walk over the ice just in time! George Lowe had a marvellous day's photography! Penguins, seals, weasel trip, icebergs, sunshine and drama! What a film they can make on returning.

The day is still perfect—I have no idea what broke up the ice! After the break up I was looking out of my porthole and saw a number of quite large swells go through, past the ship and on under the ice. I went to draw other people's attention to them to see if they presaged a storm, but they did not recur. Apparently they represented the surge from an iceberg which had cracked in half or calved or turned over! They do that in hot weather.

Gosh I'm tired! Three hours sleep the night before last, four last night, long walks and hard work each day. I did four hours work before breakfast today! I'm going to bed—it is already 2310 hours!

11

SLEDGING TO LICHEN ISLAND

The ANARE, in 1955, made the first deep penetration into Prydz Bay, south of the location of Davis Station.

No section of Australian Antarctic Territory offers as much to the explorer as the Prydz Bay region. On the eastern side is the West Ice Shelf and, further south, the remarkable rock 'oases' of the Vestfold Hills and the Larsemann Hills, together with groups of ice-free islands—the Rauers and the Bolingens. At the bottom of the bay is a jumble of glacier tongues, fast ice, icebergs and ice shelves, the great Lambert Glacier and various mountains and nunataks, such as the Munro Kerr Mountains. To the south-west and west lie the extensive Amery Ice Shelf and the edge of the Antarctic continent, running from south to north to terminate at Cape Darnley.

Access to most of these features is difficult, or rather, was difficult before the advent of helicopters. At sea, one's approach is obstructed by iceberg accumulations, heavy pack and fast ice. The interpretation of aerial photographs, particularly those covering the southern areas, is complicated by the jumbled nature of the ice forms and the ever-changing boundaries of the glacier terminations and the ice shelves.

Between the years 1954 and 1966 I directed a number of ANARE attempts to explore this region and delineate its features but, at the end of this period, I still thirsted for more information, for our investigations were far from complete. Now that Law Base has been established in the Larsemann Hills and field parties have the use of helicopters, the last remaining puzzles of this extraordinary region are finally being solved.

Early in 1955 we made our first attempt to penetrate deep into Prydz Bay. The following extracts from my diary tell the story.

Wednesday 2 February 1955
Noon
Position: 69°12.5'S 76°02E (sun) off Larsemann Hills
Wind: NW10k
Barom: 992.2
Air temp: 28.4F
Weather: Perfect, calm, clear, sunny.

We sailed south at 1616 hours for Sandefjord Bay. I got some very good icebreaking shots of the ship. We moved through some heavy floes into open water and soon reached our destination. To the south could be seen land outcrops behind massed icebergs and ahead what looked like an ice shelf with terminal cliffs about 18-20 metres high.

We pushed a short way into the fast ice edge and stopped at 2000 hours. I had a good look at it all and decided to organise a sledging party of six to cross the fast ice to reach a rock feature which I thought might be Mount Caroline Mikkelson. During a discussion which I held with Thompson, Béchervaise, Shaw, Krohn and Lacey, we examined Roscoe's rough map of the area (a cartographic compilation from US 'Highjump' aerial photographs) and it seemed to us that the Polar Record Glacier Tongue no longer existed.

I set the men packing the sledge and chose Béchervaise, Shaw, Krohn, Ward and Lacey to accompany me on the sledging trip. Then Thompson and I set off on skis at 2110 hours to reconnoitre the fast ice.

We were away until 2315 and did about seven kilometres. The surface was powder snow and we moved over it with ease. When we reached what we thought to be an ice shelf (but which proved later to be the 'missing' Polar Record Glacier Tongue) we found a long snow slope leading up onto it, so up we went. The surface was good, with occasional small crevasses a few centimetres to a metre wide, which supported our skis okay. This was the first time I had been on a glacier tongue and it was an exciting experience. There was no wind and the temperature was around 20°F. In the eerie, grey twilight of the Antarctic midnight, the sloping surface of the glacier tongue rose steadily ahead of us and disappeared in the gloom. Along its eastern edge myriads of icebergs and smaller bergy bits rose ghost-like from a blue-grey plain of fast ice while, beyond their summits, we could just discern the darker shapes of the rock outcrops far to the south which were to be our objective.

I decided it would be a reasonable 'calculated risk' to sledge over the fast ice because, immediately west of our proposed route, the glacier tongue with its occasional snow slopes for access would provide a safe refuge in case the fast ice were to break out.

When we returned, Béchervaise told me he had plotted Roscoe's map over the Norwegian Hansen chart and had taken bearings also. As a result he had determined that the ice shelf ahead *was* the Polar Record Glacier Tongue and the rocks we were looking at were the Bolingen Islands.

It was clear that if we were to sail around the glacier tongue into Sandefjord Bay we would finish up nearly 50 kilometres from Mount Caroline Mikkelson. Where we were, we were only about 20 kilometres from a 90-metre island which was quite distinguishable and would give a good reference point if we could land on it and obtain an astrofix.

We decided to do this and Fred Elliott stayed up all night packing the sledge. I got to bed at 0030 hours and asked to be called at 0500.

While Thompson and I were away skiing, a large iceberg had moved down with the southerly current, swung around north of the ship and passed westward around the point of the glacier tongue. It was reassuring to know that the icebergs follow a current which swirls off the submerged sides of the glacier tongue and this carries them safely around the corner out of the way of the *Kista Dan* Two smaller ones had followed the big one.

Whether due to the iceberg movement or the swell or the weakening of the ice by the ship's charges I do not know, but when Dick and I returned, we found that the fast ice near the ship and for about a kilometre south had broken up.

The cliff-face of the glacier tongue showed marked bands of alternate névé and ice. The firn is the result of winter snowfalls, while the ice bands are due to melting during the summer. Such annual layering is used by glaciologists when dating ice cores obtained from drill holes.

Thursday 3 February 1955

Ship approx. 69°11S, 75°34E (Dead reckoning from previous noon position) Just off Polar Record Glacier Tongue.

Wind: E 4-5 20K, then NE5

Barom: 991.1

Air temp: 27.4°F

Sea temp: 34.2°F?

I was awakened at 0400 hours by Peter Shaw, who told me that the barometer was dropping, there was a NE wind and heavy black clouds were coming up from the north and north-east. As it looked threatening, we called the captain.

I expected him to say we must get out of here, where a NE wind would drive pack ice down onto us and we were to windward of the great glacier tongue. However, he said that if there was to be a blow he would rather stay here than go into open water and that he would prefer this situation to the anchorage at Vestfold Hills. Arguments: If a blow, there will be a swell, which will break up the floes to seaward of us. The floes are soft enough to break if they push against the ship. The icebergs will not move down on us but will sweep around the corner. Anyway, danger from moving icebergs is a risk anywhere. I personally felt we would be safer in the shallow iceberg-free anchorage at Vestfold Hills, in the lee of the land for a NE blow, but did not press the point and, as he expressed himself quite happy to remain here, I agreed. If the blow were to clear out some fast ice, we might get closer to Bolingen or Larsemann; if the blow does not eventuate we will go ahead as planned.

I went back to bed and slept from 0500 until 0830. After breakfast I was busy packing and cleaning up my cabin, which had a wondrous assortment of gear cluttering it up.

As the blow did not eventuate and the day smoothed out into one of medium 20-knot wind, stable conditions and snow showers, I decided to push ahead with the scheme of visiting Bolingen Islands.

I sent Lacey out for ski instruction by Peter Shaw. Shaw found immediately that the bindings on the skis did not fit the US ski boots with which the men had been issued. (I had asked everyone for weeks past about this and they had all said they were OK. I remembered asking even in Melbourne, because I had found that the boots would not fit my bindings—that was why I had brought my own ski boots.) Delays in general packing together with adjustments to skis and bindings delayed our start until 1730 hours. (George Lowe was busy photographing us, too.) I made the boys have a meal before they left.

Eventually we got away. The sledge was heavy and pulling was far from easy. The first few miles were painfully slow, as someone always seemed to be stopping to adjust a binding or to replace a ski which had come off. The moral to this is *always* test gear before use. In this case it was not possible—how I wish we had time every year for a snow session at Mount Hotham!

The snow was not as good as when Dick and I went out the previous day. The temperature had gone up and the powder was half-damp, although not slushy. There was an equal proportion of crusty, iced snow also, over which the sledge pulling was harder. I think these crusty ridges contained crystals of salt, because on pure ice the sledge ran well.

The lack of fit in the bindings caused the boots to cut through the toestraps—the soles were too thick to fit under the welt lugs. This trouble, and skis slipping off, caused numerous delays. The sledge and sledge wheel caused more.

At the 2 km mark we went beyond the region we had surveyed on the radar and followed along the eastern edge of the glacier tongue snout. The cliffs showed crevasses, bands of ice from previous seasonal thaws and great variations in height. In some places the tongue sloped down to meet snow drifts up which one could easily ski and climb to reach its surface. Further south it seemed to me that the banded structure was less evident and that the whole thickness was blue ice without névé.

The wind was behind us, which helped considerably (NNE 20 knots). We put flags in at strategic points. Snow was falling but not thickly and, although everything was grey and dim, the visibility was not bad. Earlier it was 800 metres but later became 3-4 km. As each distinctive feature loomed up ahead we would note its bearing and sledge for it. Upon arrival we would take a back bearing on the feature we had just left.

The first phase ended when we reached a point where great bergs in tumbled profusion had piled up against the edge of the glacier tongue. Peter Shaw and I went ahead on skis to see if we could push through between the bergs. Recently we had gained the impression that we were climbing, and I was hoping that this might be so and that we would find ourselves on semi-permanent bay ice 10-15 metres thick. It was disappointing, therefore, to discover swell cracks around the icebergs and newly frozen black ice, some of which was quite slushy and tough and salty. Obviously we were still at sea level. Later it eventuated that for the whole journey we were on sea ice which, except for occasional drifts and mounds, was only about half a metre above sea level.

I did not feel very happy about the condition of the ice and prayed that no storm and swell would arise to sweep it out to sea. Shaw and I found hummocks and snowdrifts, huge tilted icebergs and piles of broken ice, pressure ridges and weak black ice. Altogether it was a fairly impassable place. Our heavy sledge was the limit of what we could pull and even small ridges taxed our strength. Whenever the sledge stopped, the runners would freeze onto the surface and we would have to count and jerk in unison to start it again, often counting and jerking to 8 or 10 times before it would budge.

From this point on I chose our route with great care, trending in the general direction of the islands but choosing open areas of reasonable smoothness, preferably between distinctive icebergs which would act as markers for us. These I drew in my book. Where any major change of route direction occurred, we put in a red flag on a bamboo pole. A vast accumulation of bergs now lay ahead and on our right as we moved out from the glacier tongue, although there were still many icebergs on the left.

We found ourselves in a veritable maze of icebergs. Their oppressive grandeur was heightened by the dim half-light. The bergs and bergy bits numbered hundreds, Immobilised in the fast ice. Most appeared to be floating, for there were no tidecracks around their bases and no tide erosion marks up their lower sides. There were, however, small surrounding cracks (in the fast ice) caused by sea swell resulting from storms.

Much of the area we traversed was weak ice, often black and newly frozen, extensively cracked, with much salty slush. There were frequent eerie sounds, like bullets ricocheting, as the miscellaneous ice systems moved under various compulsions. Pressure ridges occurred where floes had been forced together, often half-covered by drift snow. Along the eastern edge of the glacier tongue the icebergs presented an awesome spectacle. Under the influence of strong currents and violent storms these million-tonne leviathans had piled up on the immovable glacier tongue and upon each other. Their tilted convolutions and the jumbled convolution of shattered ice resulting from their violent collisions were stark evidence of the gigantic forces unleashed during Antarctic hurricanes. One dared not envisage being in this vicinity during a heavy storm.

Route finding became a matter of inspired guesswork. One would look around and see several gaps between the encircling bergs that would lead roughly in a useful direction, but which should one choose? On several occasions, as in a maze, we chose wrongly, found ourselves at a dead-end, confronted by sheer ice walls and piled ice rubble, and had to retreat and choose again.

One such occasion was at the 7-9 km mark. There were two gaps; the one we chose led through to a jumble of icebergs, so we returned and tried the alternative. It proved promising, so on we went. At that stage we contacted the ship weakly by walkie-talkie radio for the first and last time on our outward journey.

We were becoming very tired as I then set course for a very distinctive gap between two bergs ahead and prayed that it would not be choked up by piled ice or unsafe because of tide cracks. The intervening distance was very bad sea ice—cracked, newly frozen, moving with the swell, unsubstantial. As we neared the gap, Peter Shaw asked could we stop and cook a meal, as we had been going hard since 1730 hours and it was then nearly 0200 hours—more than eight hours, during which time we had drunk some lemon drink and eaten some chocolate. I was very glad I had ordered the meal before leaving the ship!

The trouble with cooking a meal is that it involves putting up a tent. The visibility was now poor, although it looked much better than it was because of the reflected light from the snow. For example, I wanted to take a photograph and found my meter did not register any light at all, yet to our eyes the scene was quite plain and bright. However, it was difficult to see things at a distance and to assess the relief of the surface over which we were travelling.

Béchervaise began to rig one tent while Peter and I skied on to see what the iceberg gap offered. Through the gap proved a better route than an alternative one behind the berg to the east, although the black ice at the gap was cracked and water appeared in a number of places. We then decided to stop, eat and sleep for four hours before proceeding, so our second tent was erected. Through the gap we could see a black, rocky, ice-free island which we assumed was Lorten, an outlying island of the Bolingen Group on the Hansen map. This boosted our morale, for it was roughly eight kilometres closer than any of the other islands and just as good for our purpose. It looked nice and close! When I was chopping ice from an iceberg for water it appeared to heave up out of the tidal crack and subside again. It was only the swell moving the sea ice, not the iceberg moving.

It was 0400 hours before we got to bed. We slept until 0800 then ate breakfast, took photos of this unrealistically beautiful camp and got under way again at 1130 hours, heading straight for Lorten. It was Friday.

We negotiated the gap all right but I could imagine it to be a burnt bridge should any swell or wind break up its weak coating of newly frozen, black, cracked ice. The memory of this gap remained a worry to niggle at my mind for the next 34 hours.

At various times we heard a noise, somewhat like the sound of an express rushing through a station, caused by ice crashing down from icebergs. After a hot day it is dangerous to go too close to them! We saw two which had, within only the last few days, suffered from falls. On each, the whole of one face, to a depth of about one metre, had flaked off like shale and lay in a pile of litter at the base. Ice is a bad conductor of heat. When hot sun shines on the vertical face of any iceberg, differential expansion between the surface and the inner layers causes fracturing.

As we approached Lorten, I could see what looked like black water around the base, with ice a metre or so higher up on the rock. In dismay I realised that we might have made all this effort only to find ourselves prevented from landing by a narrow open belt of water around the rock! We pushed on, making very good pace compared with the previous day, past silent groups of emperor penguins clustered together motionless on the snow, past numerous Weddell seals basking on the sea ice, over crunchy névé, powder snow and ice hummocks, until at last we stood half a kilometre from the island and saw that it was rimmed on the north and east by newly-frozen black ice.

A metre above this ice, clinging to the rock and projecting as a shelf, was a ledge of ice marking the upper limit of the tidal rise, and beneath this ledge was open water. On the western side an extensive pool of open water reached right to the rock of the island.

Leaving the sledge where we were, we hurriedly ate the lunch we had prepared that morning—four biscuits each with butter and marmite, an orange, some chocolate, and hot cocoa from a thermos—and pushed over to Lorten on skis. We found it most interesting, with a wealth of lichens, great expanses of moss, interesting rocks, and excellent view points from which to survey our surroundings. Krohn collected and examined rocks, Lacey set up his theodolite and waited for a sun shot, I took panoramic photos and Béchervaise collected lichens and mosses. Jack Ward built a rock cairn and attended our radio. We noted a swell at the ice crack of about 10 to 20 centimetres.

We returned to our sledge on the sea ice at about 1800 hours, had a successful radio schedule with the ship, pitched our two tents on the highest surface I could find, and cooked dinner. Again I was not happy about the camp site. I did not wish to go onto the island in case the black ice melted or broke up and left us sitting there stranded and waiting for the next freeze. On the other hand, there were no heavy floes anywhere handy which we could camp on—the best I could find was a spot about a metre higher than the uniform sea-ice surface, which apparently was a pressure ridge covered by snow.

The snow had firnified or crystallised into ice, giving under us about a metre of granular ice like loose gravel. This overlaid the usual sea ice, which did not appear to be more than half a metre above the water (and probably one to one and a half beneath).

The evening meal was pemmican 'hooch' (pemmican, potato powder, onion powder, oatmeal) one biscuit each, coffee, chocolate. I then sat and wrote by the light of an electric lamp, and later Béchervaise came in and produced a couple of cigars, which we smoked. Most relaxing! There was a radio schedule at 2200 hours, at which we arranged a listening schedule for 0600 next morning and a speaking schedule for 0900. Thompson sent us a warning message that there was a one and a half metre swell at the ship, which had broken up the ice for some distance along our track. We went to bed shortly after 2300 hours in what looked like 'set fair' weather. The air temperature was 20°F.

I was worried about several things. First, a hurricane would set this area into a heaving, gigantic cauldron in which icebergs would roll and grind to slush all the surrounding pack that was not broken up by wind and swell. In such a case our only safety would lie on the island. Second, I was concerned that our way of return might be blocked at the two weak points—the Gap and the rotten kilometre of ice the other side of it. A 3-metre lane of water at the gap could prevent our escape and force us to search the iceberg maze for another route under conditions where time might be paramount.

Third, I was concerned (but less so) as to what we would do if wide lanes of open water prevented our reaching the ship and the heavy floes prevented the ship from reaching us. This would not be serious in good weather, as we would simply wait for a re-freeze or a close-up. Fourth, there was the worry that a storm might break up all the sea ice and disperse it in conditions of bad visibility. In such a case we would drift out to sea on an ice floe without the ship seeing us.

With these thoughts on my mind I went to sleep. At 0300 hours I was awakened when Bob Lacey got up and went outside. He returned and said, 'There's a wind getting up, Phil.' I said, 'What direction?' and he replied 'South, I think.' So I said, 'Fine,' and turned over. In spite of my warm bag, however, I was not able to leave it at that and so crawled out and looked for myself.

It was north-east, not south, and snowing. The snow was driving along almost horizontally and the tent was flapping. I thought, 'This is what I've been waiting for—this is what broke up the ice near the ship.' I awakened Peter Shaw, who said he thought the weather looked bad. His aneroid barometer had fallen, but not greatly. 'Right boys, get moving,' I said. 'We're going to shift camp onto the island.' We worked like demons, without breakfast, and took from 0300 hours until 0430 hours to pack our tents and gear on the sledge and set off. We hauled the sledge over the weak black ice and stopped beside the tide-crack ice ledge. There we unpacked the sledge and started to pass things up to one man on the island.

I decided to get up there and help him, so kicked off my skis and walked to the edge of the black ice, where it melted into the pool beneath the ice ledge. I did not like the look of it much, so took an ice axe and dug it into the ice ledge, grabbed hold of it and sprang. The spring did it—the ice broke, the ice axe came out and I went straight down into the water. However, I was lucky—there was a second ledge a metre down and I bounced back off that onto the ice quicker than you can count. But I was wet from the thighs down and both my cameras (slung around my shoulders) had been immersed. We therefore hurriedly rigged our tents on the island. In one I lit a primus stove and changed my wet clothes, then spent time drying out the cameras. While the others continued to fix the camp, I cooked breakfast. We were comfortably established ashore by 0600 hours.

After breakfast I busied myself cleaning up the tent and drying my wet clothes, then wrote notes on the journey and the island. Lacey took a sun shot at 1030 hours and others approaching noon. The night before I had written a note about our landing, so we placed this in a rum bottle and put it beneath the cairn. We completed our collections and then abandoned any further astrofix observations as cloud obscured the sun. We had been lucky to get the noon shot! I decided then to make a large meal instead of lunch, to pack immediately afterwards and to dash for the ship.

Our island had proved to be quite fascinating. Although it boasted no penguin rookery, it apparently was used as a roosting place by moulting Adélie penguins, whose guano had stimulated a prolific growth of lichens, algae and mosses. The brightly coloured beds of lichens, of a variety of colours, were particularly impressive. I decided to replace the Norwegian name 'Lorten' with the name 'Lichen Island' aesthetically more acceptable than Lorten which, translated, means 'The Turd'. During our sledge journey we had seen numerous MacCormick skuas and Wilson's petrels. At intervals we had passed groups of Emperor penguins, 14 to 20 in each group, standing motionless as though waiting for a bus. There must have been an Emperor penguin rookery further south of the island. In the early stages there were a few snow petrels and on the island I saw a giant petrel alight on a seal carcass and drive off the skuas

There were many Weddell seals on the ice around the island. We counted 50 on one long stretch of ice fringing the large pool west of the island. Lacey fixed the position of Lichen Island as latitude 69°20'S, longitude 75°32'E. The magnetic variation was found to be 71°W. The island was free from erratic boulders and glacial debris, and the country rock consisted largely of biotite para-gneisses grading locally into biotite hornfelses.

The island was elongated in the direction ENE-WSW and was roughly dumbbell-shaped, with knobbly ends connected by a low saddle. It measured roughly 315 metres long and 135 metres wide. It can be considered to be a north-westerly outlier of the Bolingen Islands, which lie off the Ingred Christensen Coast between the Larsemann Hills and Sandefjord Bay.

We had our meal at 1400 hours. By the time we had packed and were ready to leave it was 1730. The day had not turned out badly—sunny in the morning, then the wind disappeared, it became relatively warm and the snow began to thaw. The thaw plus the slight swell made me worry still about the condition of the Gap, but the fine weather took the urgency away from all our activities.

We had fairly easy hauling over the thawing snow to the Gap which, to our relief, was still negotiable, although a number of fresh cracks had developed. Beyond the Gap, to our delight, we saw the *Kista Dan* silhouetted against a sunlit iceberg straight ahead over a clear expanse of sea ice. She had moved from her anchorage and broken in six kilometres towards us through the breaking-up fast ice. We would therefore not have to thread our way back the way we had come through the iceberg maze!

It was now hot work hauling, as the wind had completely gone, but one became cold very quickly upon stopping. We went along very well, but the hauling became harder as the evening wore on and the snow became drier. We abandoned our old track and struck out directly for the ship. A glorious sunset developed as we proceeded and we stopped for a few moments to photograph it.

The last five kilometres of the ice towards the ship had deteriorated. First there were small cracks, then wider ones, finally ones 1-½ metres closed with the swell. The ice was half a metre above the water and one metre below, all blue, hard ice except for yesterday's snow. One large crack three metres wide forced us to detour 400 metres and several 1.5 metre cracks also forced appreciable deviations. More generally, though, we would wait on a floe until it drifted into contact with another and then would haul the sledge across.

The captain radioed us that he had rammed the bows of his ship into the side of a large ice floe and was pushing it hard against the other ice to prevent it dispersing. We were able to continue our floe-hopping progress until we reached the large floe at 2245 hours. We then had to wait until 2315 while the ship backed off and manoeuvred to come alongside.

By then we were cold, tired and hungry and greatly relished a hot meal and shower on board. I got to bed about 0130 hours. There I reflected upon our good luck. If there had been bad visibility—fog or drift snow or blizzard—we would not have been able to see the ship and would have drifted out to sea with the floes.

12

THE FIRST AUSTRALIA–ANTARCTICA FLIGHT

It was not until 1964 that the last intercontinental flight remaining in the world was accomplished, that from Australia to Antarctica.

'Hang on, we're going in,' called Admiral Reedy over the intercom from the flight deck of the Hercules. Eleven of us tensed as we sat on the metal seats arranged down the sides of the cargo space at the rear of the aircraft. Seven others occupied the flight deck, above and forward of where we were. Between us, occupying most of the cargo space, was a huge cylindrical tank holding 16 000 litres of fuel for our long flight.

We had been sitting up forward in the fuselage, in front of this tank, but James Reedy had said, as he issued each of us with a crash helmet, 'If we hit anything, this tank will take off like a projectile and crash through the front of the plane. Up forward on the flight deck we have to risk that, but there is no reason why you others should.'

There were no windows where we sat. We had no way of estimating our descent or of knowing how close we were to landing. We could only sit and wait—and hope!

Several minutes dragged past as our hearts pumped faster and our breathing quickened. Then, over the roar of the engines, we heard a sizzling noise and realised the plane's skis were running over snow. Another fifteen seconds passed, the bumping of the plane on the surface slowed, then the front of the aircraft lurched downwards and we sprawled sideways as the plane slewed around in a series of swirls and came to rest.

As we stepped out onto the firm snow of the Antarctic plateau, James Reedy said to me, 'You can always tell when men have been under stress. They talk and laugh too much!'

Some 15 hours earlier we had left Avalon Airport (between Melbourne and Geelong). Historically, our flight would be significant as the last pioneer intercontinental flight in the world. Although a US aircraft had, in 1960, flown from McMurdo to Hobart, Australia, no flight had ever been made from Australia to the Antarctic continent.

After a number of years of Antarctic experience I had, in Melbourne, approached this journey as a pretty routine exercise, appreciating the opportunity to take part in an historic venture and looking forward to seeing the South Pole, but expecting little in the way of excitement.

Planned and directed by Rear Admiral James Reedy, United States Navy, Commander of the US Support Force in Antarctica, it was proposed that the Hercules LC130-F aircraft should fly to Amundsen-Scott Station at the South Pole, land there and deliver mail and fresh food, then fly on to McMurdo Station on Ross Island in the Ross Sea. However, our flight was to turn out to be anything but routine.

The Hercules left Avalon Airport at sunset on 30 September 1964, with 18 men on board. Captain of the aircraft was Commander Fred S. Gallup, commander of Operation Deep Freeze Squadron VX6. With a crew of six and Admiral Reedy, there were eight on the flight deck. Seven other USN crewmen travelled in the passenger accommodation, together with the admiral's chief meteorological adviser (Commander Ken Allison, USN), Australian journalist David Burke and myself. Total take-off weight of the aircraft was around 60 000 kilos. The whole flight was navigated by astronomical means (star and sun observations) without the use of beacons of any sort, except for radio compass on nearing our objectives.

The take-off from Avalon was uneventful, with my wife Nel waving from the tarmac. By the time we passed the coast of Tasmania it was dark and those with no duties sought sleep in the bunks fixed to the walls of the fuselage. Engine noise inside the plane was deafening. I tried the earplugs provided, but they were uncomfortable and made little difference to the noise. I lay on the comfortable bunk, pulled up a blanket, and tried to ignore the throbbing din.

David Burke shook me at 2100 hours to tell me that an aurora was visible from the flight deck. The night was clear and there was no moon. The stars shone with unusual brilliance and, stretching from the horizon on the port side, up and across the sky ahead of the plane, there shimmered a broad, bright, greenish-white band of light.

Many Antarctic visitors have never seen an aurora. The best and most frequent displays are confined to a narrow annular ribbon of latitude which encircles not the geographic pole but the geomagnetic pole. If the station at which one lives is far from the 'auroral zone' and the sky is frequently overcast, one may pass a whole year without seeing a good aurora. If one is a summer 'tourist' continuous daylight precludes any chance of a visible display.

This aurora, although bright, was disappointingly inactive, shining placidly with little movement and none of the vertical rays which, twisting and weaving, provide in the most spectacular displays the exciting effect of a waving curtain of colour in the sky. However, to me this aurora was unique, for instead of being earth-bound, I was flying towards it at 450 kph.

I wanted to see the Antarctic coast as we crossed it, so at midnight I rose again and, from the flight deck, peered out. Far below, faintly visible in the starlight, I could just discern large ice floes and, occasionally, large icebergs separated by black lanes of water. While I concentrated on the dim scene below, the navigating officers were concerned only with the sky. Using astronavigation and plotting on a grid system, they were aiming to cross directly over the South Magnetic Pole which then lay in the region of the Mertz Glacier Tongue, at 145°E longitude.

At 20 minutes past midnight I saw the edge of the Antarctic continent slide below us. It trended away to the west with the broken black and white pattern of pack ice on the north side of it and nothing but blank whiteness to the south. We had left Avalon at 1756 hours, so our first 1 790 nautical miles had taken 6 hours 24 minutes. My last journey across the same expanse of ocean had been in a polar ship on which I had spent days of misery, rolling and pitching in the stormy seas.

At 0300 hours the sun rose over the distant mountains fringing the southern end of the Ross Ice Shelf. Below us, the surface of the Antarctic Plateau stretched as a vast and featureless desert, changing from dull grey to faint rose as the sun's slanting rays brushed its surface.

As we flew on at 10 000 metres towards the Pole, a large window at floor level on the port side of the flight deck suddenly shattered like a windscreen of a car when struck by a pebble. Fortunately it remained in place and did not blow out or we should have lost our pressure and been forced to dive rapidly. I did not know what caused this.

At 0555 hours, the Hercules, which had been descending for some time on a long, gentle glide, banked sharply and we knew we were over the Pole. We circled once at about 2 900 metres (some 100 metres above the plateau surface), while crew members slid up the large door on the port side for bags of mail to be dropped to the party at the Pole Station. The temperature on the surface was -85°F, which was far too cold for the Hercules to land. We came around in a wide arc, then swept in on a long straight run low over the station. The bags were thrown out to the lonely-looking figures below, the crew began closing the door and the plane climbed steeply to regain its economical cruising altitude as it turned north for McMurdo. No aircraft had ever flown over the Pole so early in the Spring.

The crew struggled with the door, which stuck obstinately as the aircraft climbed past 6 000 metres. Apparently the cold polar air, entering the open door, had condensed vapour from the warm interior of the plane and this, running down the sides of the door, had frozen and was preventing the door from closing over the last 20 or 30 centimetres. We found ourselves flying unpressurised at an altitude of over 6 000 metres.

This was a most uncomfortable experience. The main sensations were breathlessness, nausea, some giddiness, lack of muscular coordination and an oppressive feeling of claustrophobia.

We were asked to sit in a group and an oxygen mask was passed around as the pilot brought the aircraft down to around 4 000 metres. It was like a school of opium smokers or a 'pot' group; one would have a few sucks of oxygen, then pass the gas bottle on to the next man. By the time it returned, one was again breathless. We could not descend lower then 4 000 metres because there were still mountain ranges to be crossed on the way to McMurdo. Then we received a radio message from McMurdo stating that the weather had closed in, a blizzard was developing and a landing would not be possible.

Our alternative airfield was at Byrd Station, in the middle of nowhere, deep in the interior of the continent in Marie Byrd Land. We headed for Byrd at 0635 hours and, flying past Mount Johansen, we rose to 5 000 metres, again taking turns to sniff oxygen.

The scenery below was magnificent. The warm, golden, early light bathed the cream snow and dark-brown rock cliffs of the Queen Maud Mountains. Through gaps between the peaks tumbled the terminal ice falls of glaciers, as the plateau ice sheet, dammed back behind these rock ramparts, burst through to surge down towards the Ross Ice Shelf.

The crevassed streams of the Robert Scott and Leverett glaciers, like turgid currents snap-frozen into immobility, wound down through steep-walled valleys beneath bare, jagged peaks and snow-filled cirques, to where the placid expanse of flat ice shelf stretched monotonously to the horizon.

I photographed madly for 10 minutes, short of breath in the rare, unpressurised atmosphere, trying to record as much as possible of this superb scene before it disappeared behind us. Then we were out over the featureless Rockefeller Plateau, and there was little of further interest until we arrived over Byrd Station at 0810 hours, Melbourne time, on 1 October. The plateau altitude here was 4 500 metres and the Hercules began circling the station. We had come 7 100 kilometres non-stop to this outpost at 80°S latitude and, when we landed, our flying time of 15 hours 39 minutes was the longest duration of any flight in Antarctic aviation history.

After some five minutes of circling, with no sign of our landing, I went to the flight deck and asked James Reedy what was delaying us. He said 'We can't get the nose ski down' The crew tried various expedients without success. They then removed some floorboards and drew out a heavy chain that was secured to a lever mechanism connected to the nose ski. Ten of us heaved on the chain, like a tug-of-war team, jerking it to try and force the nose ski down, but to no avail.

At the end of an hour's circling our fuel was dangerously low, so Commander Gallup decided to land without the nose ski in position. For a few minutes there was frantic activity. All hands hurriedly shifted a high stack of kitbags and suitcases from the front to the rear of the plane. Finally, passengers and all crewmen not required on the flight deck were issued with crash helmets, and strapped to seats at the rear of the fuselage. I, for one, was glad to be sitting now behind instead of in front of the large fuel tank. Then the commander said, 'Hold on, we're going in.'

It was 0935 hours when we came to rest at the end of the compressed-snow runway. It was bitterly cold outside on the snow as we walked about a kilometre to the station. There was a 20-knot wind and the temperature was -40°F.

The entrance to Byrd Station was a gaping cavern at the beginning of a long descending ramp of snow, for the whole establishment was buried under some four to five metres of ice. The station had been constructed by cutting deep long trenches in the névé snow of the plateau and then building the various huts positioned centrally in these trenches. The whole excavation had then been roofed with arches of corrugated iron so that the top lay level with the surrounding snow surface. In this fashion the designers had hoped to avoid the problems of 'drifting over' by snow that buildings erected on the plateau surface experience.

Just inside the cavernous entrance was the vehicle garage and workshop and here the diesel fumes of running engines, combined with the very low temperature of the air, gave me a violent fit of asthma—something I had never before experienced. Wheezing and choking for breath, I hurried through to the purer air of the buried passageways and huts beyond.

The men at this station lived like troglodytes, for the most part unaware of the passage of day and night. Their movements were governed by clock time, and they lived in perpetual electric light. The ice of the tunnel walls looked like fine plaster and was dry and rough like concrete to the touch. Its temperature never rose above zero Fahrenheit. When a hut door was opened a white cloud of condensed vapour billowed out as though the interior were on fire.

But they were having trouble from the constant accumulation of precipitated snow on the plateau surface and the slow glacier-like movement of the ice sheet (here 2 500 metres thick) towards the coast. The result was a twofold creep of the ice; a subsidence, a compression in the vertical direction, and sideways creep which caused the walls of any pit gradually to encroach. Subsidence was the greater problem. In many places, water pipes and electric cable attached to the walls were already sagging dangerously. A few days before we arrived a water main had snapped because of subsidence. Nevertheless, the trench system had proved greatly superior to the building system of the old Byrd station. Built on the surface and allowed to be covered by drift snow, the station became so distorted by ice pressure that it had to be abandoned after six years.

I never managed to sort out the time and the days in this warren of Byrd Station. McMurdo time was two hours ahead of Melbourne time, but, because it was across the dateline, Byrd was 18 hours behind McMurdo. In the village of electric light beneath the snow, out of touch with daylight and darkness and working from 12-hour clocks, I became hopelessly confused between breakfast, lunch and dinner times, and rising from my bunk after what I thought had been a two-hour nap, I found to my disgust that I had wasted 14 hours in deep sleep.

Next morning we walked out to the airstrip in time to see a second Hercules arrive from McMurdo with fuel for the admiral's plane. McMurdo airport was again open to traffic. Working all night, mechanics had repaired the hydraulic gear of our plane's nose ski, whose failure had been caused by ice crystals clogging up a valve. It was explained to me that the hydraulics had recently been completely overhauled in the USA and apparently, by mistake, the pressure system had been recharged with compressed air instead of compressed nitrogen. Water vapour in the air had condensed and then frozen in the valve.

Admiral Reedy insisted that Burke and I travel to McMurdo in the second aircraft, just in case his own suffered further ski trouble. Finally we took off in the wake of the admiral's Hercules. After our cold wait at the airstrip we did not warm up for a long time.

Dusk was creeping in as we approached McMurdo, with Mount Erebus and Mount Terror rising majestically on our starboard side. We looked down into the steaming, rock-lipped crater of Erebus and I thought of young Douglas Mawson who, with Australian Edgeworth David and several others from Shackleton's expedition, made the first ascent of this mountain in March 1908.

It was not as cold at McMurdo as at Byrd. At the station, 4.5 kilometres from the runway, across thick semi-permanent sea ice, I learned that a Hercules was due to take off in three hours' time for New Zealand, so I decided to accompany it. After a quick drink with the admiral and his staff, an excellent dinner in the well-appointed mess and a rapid look over the nuclear energy plant, which provided all power requirements at McMurdo, I was back at the airfield, soon to be airborne and flying north past Mount Erebus.

Fifteen minutes after our departure the pilot announced that the radio had failed and he was returning to McMurdo. When we landed he suggested that, as he did not know how long the fault would take to remedy, we should stay conveniently near the aircraft. Accordingly, the dozen passengers stumbled, in the teeth of a cold wind, towards the lights of the main passenger terminal, an oil-heated Jamesway hut, about 270 metres away. The front section was an office, the back part a dormitory with two-decker bunks. Someone found us a blanket each, so we threw ourselves down in our clothes and were soon asleep.

Next morning, we trudged across to the mess of the squadron that ran the airfield and, after a hasty breakfast, returned to the aircraft. At

0708 hours (Melbourne time) we took off again for Christchurch and this time our flight proceeded without incident.

There is no mountain scenery in the world to excel that of the western coast of the Ross Sea. The morning was fine and clear and the great ramparts of the Prince Albert Mountains extended back, tier upon tier, valley behind valley, into Victoria Land. American surveyors in the last few years with tellurometers and theodolites have frog-hopped their way from peak to peak by helicopter to map this complex confusion of ranges.

Farther north we crossed the coast, over the flat floating plain of ice called the Lady Newnes Ice Shelf, and passed on across the Victoria Mountains and the Tucker Glacier. To the west lay Oates Land, whose north coast I had explored from the sea in 1962. There, in the distance, I could recognise the great Rennick Glacier, running south as a broad white expanse of ice which, far inland, merged with the extensive névé of the high plateau. Only in the last three years before our flight had this region been mapped. For me this was the most interesting of all the scenery we had passed, for from our great height I had a magnificent view of the features of the interior of Oates Land that the limited range of our helicopters in 1962 had prevented. At 0330 hours (Melbourne time) we landed at Christchurch and two hours later I was in a TEAL Electra bound for Melbourne. Then came the exquisite pleasure of a shave, a wash and a fine five-course dinner in the relaxed atmosphere of the aircraft. When we arrived at Melbourne at 2230 hours, my time sense was utterly confused. I had expected to be away anywhere from five to ten days and here I was, home again, after 12 000 kilometres of eventful, exciting travel in a little over three days.

Passing through customs at Melbourne Airport I was asked to open my luggage. I was still dressed in my crumpled windproofs and my luggage consisted of a bulging 'parachute bag' and an airways carrybag (for my camera and accessories). I said 'What do you expect to find? I haven't been anywhere except the Antarctic. I didn't even leave the airport at Christchurch' But the customs man regarded me with deep suspicion and doggedly searched through my jumble of soiled socks and underwear, pullovers, balaclava and mittens. In his frustration he deliberately made the process as slow as possible. Little I cared—I was safely home again after an historic flight and a unique adventure

13

BESET

In February 1956, Kista Dan *was beset off the coast of Queen Mary Land, Antarctica.*

After travelling to Antarctica in 1950 as Australian Observer on the ship *Norsel*, I returned with the ship to Cape Town. The *Norsel* had travelled 3 680 kilometres through pack ice and I had gained invaluable experience observing a highly experienced Norwegian ice captain handling his ship.

One of his techniques was the use of explosives to free the ship under certain circumstances. Contrary to common belief, one cannot use explosives to blast a channel through a horizontal ice floe—the charges mostly just blow holes in the ice without cracking it. Should a floe be tilted, as described below, it can often be cracked if the explosives are properly placed.

When charging ice at speed, a modern ship relies upon the shape of the cutaway bow, which reduces the impact and allows the fore-part of the ship to ride up smoothly onto the ice. The weight of the ship is then generally sufficient to crack the floe and allow the ship to sink down again. With engines hard astern, the vessel then withdraws from the ice and the procedure is repeated. However, on the way home from landing an expedition, a ship may be too light up for'ard to break the ice, and may then find itself high and dry and unable to back off.

Another reason for becoming wedged when charging is as follows. If a ship charges pack ice, when there is no open water into which the ice can be pushed, there are only four places the ice can go. It can be 'rafted' out over the top or pushed down underneath the adjacent ice. It can be piled up and wedged against the ship's side or it can rise up on edge then slide down beneath the ship's keel. Ice beneath and beside the ship may wedge her so tightly that only explosives can free her.

A quantity of gelignite, varying from two to six sticks, is fitted with a detonator, fused and tied to the end of a bamboo pole. A hole is made in the ice with a crowbar two to three metres from the ship's side and in a position where the wedging appears worst. The engines are set going in the forward direction and, by the use of poles, as much ice as possible is pushed from around the ship to the stern, where it is carried away in the wake.

The engines are then set full-speed astern, the fuse is lit and the bamboo is thrust down the hole and well under the ship. A few seconds later there is a dull boom and the ship shudders as though from a 'near miss'. The explosion may shatter the ice or it may merely produce a surge and a lift of the ship. In either case the chances are that the straining propeller will drag the ship clear. If not, poles are again used to clear loosened ice from around the ship and the process is repeated. Sometimes three or four charges are exploded simultaneously along one side of the ship or even on both sides.

In February 1956, in the ship *Kista Dan* (Captain Hans Christian Petersen),we were beset for 10 days just after leaving the Russian station, Mirny, en route to Mawson.

The ship had been forced to stop at 1530 hours on 1 February in a blizzard, with winds up to 50 knots, when we ran into pack ice. By 1800 hours the pack had drifted in tightly all around us. The bad weather and poor visibility continued on 2 and 3 February but on the afternoon of the 4th the wind dropped to a breeze. The captain told me he would wait for the pressure to ease. We could see clear water only about three kilometres away, but this would be a polynia (an extensive pool), not the open sea. I noted in my diary, 'If the captain would only try the ice instead of just looking at it!' At about 1630 hours I asked the captain 'How about giving it a try?' but he said, 'No, no chance.'

On 5 February at 0600 hours, in wind of about 10 knots, the captain began backing and charging to try to turn the ship around, which he succeeded in doing by 0720. At 1000 hours, having made a distance of only 100 yards, he gave up. Our protracted delay made me fear that my plans to visit the Vestfold Hills might have to be abandoned. Next day, the weather was again bad, with winds gusting to 50 knots and visibility limited to 100 metres. The enforced inaction was very frustrating.

The day after, I noted that the ice conditions had improved considerably, although the wind was 40 knots and it was snowing heavily during the morning. The open pool was now only about 1 500 metres away.

I received a radiogram from Captain Vietrov, the master of the Russian ship *Lena*, saying that he was out at sea in a position north-west of us and asking could he help us out of the ice. I replied thanking him but saying that I thought we could manage by ourselves once the storm abated. He replied, 'Way to north-west had 60-80 per cent ice. Going through it is enough difficult because ice by storm is pressed into band approximately 35 km. I wish you happy way from ice.'

Nothing happened on 8 February, by which time we had been stuck for a week. One needs a special brand of patience for this work!

Next morning I awoke at 0450 hours to the sound of the engines. I pulled a duffel coat over my pyjamas and went up to the bridge, where I found the captain battering away at a large obstinate floe and getting nowhere, so I went back to bed. When I woke again it was 0745 and I found we had done about 50 metres; however, the captain was still trying, trying very hard.

By 1130 hours the ship had managed 100 metres but was blocked by two large, heavy floes, one of which the ship had charged a number of times, forward, then back, then forward again, riding up on it then sliding back off it, without succeeding in cracking it. My diary records: 'I cannot help feeling that if the captain had tried as hard as this on 4 February he could have got out—the ice wasn't as bad then.' Later I wrote 'It is 2015 and the captain is still at it. He took only two half-hour breaks all day, for meals, and has kept the ship working all the rest of the time. At least he seems to have awakened to some sense of urgency.'

Next day, I woke at 0500 hours and felt the ship move, so went to sleep until 0730. Going up on deck, I found sailors out on the ice with steel slings hooked to the main winches trying to haul out pieces of floe jammed beneath the ship. There was the usual mess-up—sailors putting a sling around a corner of a floe and watching it slip off when the tension came on. Why don't they cut a notch in the ice first? I went down to breakfast and, after a few more tries, the sailors gave up and came back on board.

When I returned to the bridge the captain said he thought the ship was held by floes amidships, so I offered to try some explosives to free it. To my great surprise he agreed. He really must have been desperate, for he had never agreed at any previous time on any voyage. (I can understand this—the Norwegian technique is pretty unnerving and one has to have experienced it to know that it won't damage the hull.)

Generally, in the training period in Melbourne certain men are selected to learn how to handle explosives, which are needed for occasional tasks at Antarctic stations. I had watched several of these sessions without taking particular notice, for I did not envisage ever having to handle them myself.

Now I asked my men which of them were the explosives experts and no-one moved. To my dismay not one of them had done the course. And I had made my offer to the captain!

I hate explosives. I am afraid of them. I have always kept well away when around anyone who is handling them. But there was no way I could get out of this!

I went down on the ice with the American observer, an old Byrd Antarctic man, Ike Schlossbach, and we looked all around the ship. We agreed that the ship's bows were noticeably raised and concluded that a floe must be jammed beneath the bows, not amidships. We collected Ayres and Bathurst and several other men, asked Thompson to pick up the explosives and accessories from the ship's magazines, together with several crowbars, and went out onto the ice near the port bow. I decided where to place a couple of charges and the men poked holes with the crowbars, helped by a crack in part of the floe. I then busied myself, in considerable trepidation, tying the sticks of explosive to long bamboos and affixing the detonators and wires connected to the battery and the electric plunger. All this took quite sometime.

Meanwhile the captain, on the bridge, was telling all around him that we were wasting our time, that the ship was stuck amidships and not up for'ard, that explosives wouldn't do any good anyway, and that we were a lot of fools mucking about on the ice.

When all was prepared I went to the bridge and asked the captain to signal the engines full speed astern. When the propeller was churning madly and one could feel the ship straining, I signalled the man on the plunger, he pushed it down, there was a huge concussion and the ship reared like a frightened horse. Then it slid back quietly and smoothly into the small open pool created by the wash astern.

The captain stared in amazement, then said, simply, 'Jesus Christ!'

It was now obvious that the ship had run up on a floe and cracked it so that the sections had tilted and slid down under the hull. When the ship attempted to back off it had only half-succeeded, with sections of the floe jammed down under her bows.

To minimise a recurrence, the captain now filled the forward water tanks and empty oil tanks with sea water to increase the weight of the forward portion of the ship.

We progressed only 60 metres in the next four hours. I climbed down onto the pack and obtained some good cine shots of the ship breaking towards me through great lumps of ice. At 1600 hours we still had 25 kilometres to go to the open ocean, and the ship was having a hard time trying to break through between two heavy old floes rearing 3.5 metres above the water with only about a ship's width between.

Then we became wedged again—badly wedged—right in the middle of the gap. The captain asked me to set off explosives at a certain place and, although I considered it far too wide of the ship and amidships instead of for'ard, I obliged him and we exploded three sticks of jelly without effect.

We then went up for'ard, where ice was badly wedged under the hull, and set off another three sticks, but still with no result. So, as a last resort, we tried a primitive attack. I got my expedition men out onto the ice with picks, axes and shovels and we dug ice away from the starboard side, then a little from port. Meanwhile the crew ran out a cable from the after winch to an ice anchor that they set in a floe astern. After nearly two hours work the ship was winched out backwards and floated free. At 1900 hours the captain was clearing the ice debris and manoeuvring the ship through the gap. He stopped for the night at 2400 hours.

I could not sleep at first and later I dreamed heavily, so woke far from refreshed. I was beginning to become very anxious as the days slipped by and we made no progress.

At 0500 hours on 11 February I looked out, but we were not moving. The same again at 0630. I went up on deck and found the captain out on a floe directing men fixing an ice anchor off our starboard bow. He was trying, with a cable attached to the winch astern, to drag the floe backwards, out of his way. He told me he had no room to back and charge.

They managed to shift the floe about six metres before they were forced to give up and, at 1715 hours, the captain began backing and charging in the confined space that was available—only 10-12 metres. I felt quite desperate. It seemed such a feeble, futile attempt against the great forces that the ice was marshalling against us; yet I knew that we had only one choice, namely, to push on stubbornly with our little attempts in the hope that the sum total over hours—or days—might amount to something.

Each time we backed we came hard up against the heavy lumps of ice astern, with a shuddering jar; each time we charged we advanced one or two metres. The broken ice ahead, where the ship had smashed it, withdrew reluctantly as the ship charged, then slid comfortably back into place as we backed off.

Only a kilometre of ice to open water! Yet this could wreck our plans for 1956. This trip has been one of mixed frustration and achievements, as I suppose most Antarctic voyages are. The two greatest frustrations have been from small widths of ice—at longitude 119°E, when we failed to push through to Chick Island, it was only a couple of hundred metres of ice, near the large iceberg, that thwarted us, and here it is a kilometre. If, during the next couple of hours, the captain can carve out some working room, we will be back in the race again.

Yes, at 0915 we were moving. Once past the floe that was hampering us on the port bow the ice was a bit softer and we could carve through it with better progress than for a long while. However, there were lots of hard, old, blue ice floes here and there that did not break and acted as reinforcement for the weaker ice in between.

Our icebreaking involved cutting a lane twice the width of the ship, so that water would be available at the sides of the ship into which the broken ice could be pushed. This involved repeated charges in several directions instead of the usual straight back-and-forth action. Still, we were getting somewhere.

The last 400 metres to open water were in some ways the toughest of all. The ice was mostly pulverised brash, interspersed with a few large floes and bergy bits. As the ship pressed forward, the ice closed in after her. A swell from the open ocean caused the whole system to rise and fall rhythmically when the captain charged. The friction of the 'concrete mix' brash against the ship's hull prevented him from gaining any speed and, often, the charge would not reach as far as the previous one.

About 1600 hours the tide changed, the ice closed up and a drift from south to north became evident. Some big tough floes that we had passed drifted across astern and interfered with our backing. The wind and swell were dead ahead and the brash ice, closing in like an enveloping glove around the ship, added to the resistance to motion. The captain changed his procedure to backing only 10 metres before coming forward again, thus preventing too much ice returning to block his path after the ship backed off. Some of the time, with the engines pulsing 'full ahead' the ship inched along almost imperceptibly.

I waited, on and off, with my cine camera so that I could film the supreme movement when we broke free into open water but then missed the most graphic part because the film ran out.

As we approached the last few metres to freedom, we suddenly found ourselves in a fantastic jumble of heaving ice. The whole of the pack was rising and falling some five to ten metres in the swell from the open ocean; great heavy floes and bergy bits humped themselves heavily up out of the water, then slumped back almost out of sight beneath the brash of pulverised ice, only to heave themselves out again in ugly fashion, grinding and bumping against themselves, with cascades of water pouring from their rounded shoulders and the spray flying.

The ship charged through this maelstrom and our men, crowding the railings, savoured the excitement and the danger and held their breaths as these monsters crunched threateningly only a few metres each side of the ship.

We reached open water at 1950 hours on 11 February, having been held in the ice since 1530 hours on 1 February. The captain set the ship on course for Mawson Station.

When we were at Mawson I received, on 2 March, a long radiogram from the Russians at Mirny which read:

From the press I have recognised that Kista Dan not yet arrived Station Mawson on account of heavy ice conditions and ship being squeezed by heavy ice stop If that is true I am ready to render you all disposable help stop It can be ice air reconnaissance as well as help of our expedition ship Ob which is cruising in the Davis Sea carrying out operations stop The help will be free of charge in sign of friendship between our expedition stop I am waiting for your answer.

I replied:

Thank you for your very generous offer stop I must apologise for not informing you that we had escaped from the ice but I assumed our later press messages would have been broadcast stop We reached open water SW of Drygalski Island on 11 February after three days struggle and had no difficulty proceeding from there NW to open ocean stop Arrived Mawson 17 and have now finished unloading and building five huts stop Have also erected frame of our large aircraft hanger stop We hope to leave for Australia next Monday stop With our appreciation of your thoughtfulness and with best wish for a successful year.

Hercules aircraft at Byrd Station. *(P.G. Law)*

Phillip Law with Rear Admiral James Reedy upon arrival at Williams Airfield, McMurdo, 1964. *(USN)*

After the storm — *Kista Dan. (P.G. Law)*

Horseshoe Harbour and Mawson Station, 1960. *(P.G. Law)*

Phillip Law on fast ice outside Horseshoe Harbour after the first landing to establish Mawson Station. The polar bear mittens once belonged to Sir Hubert Wilkins. *(D. Leckie)*

Ayres on plateau at top of ice slope. *(P.G. Law)*

Phillip Law scuba diving.

Glacier bergs with heavily crevassed surfaces. *(P.G. Law)*

Below left: Raising the Australian flag near Magga Peak, 21 February 1959. Left to right: Robert Dalton, Bruce Coombs, George Smith and Phillip Law. (*A. Campbell-Drury) Below right:* Victoria Land, on the Western side of the Ross Sea. *(P.G. Law)*

Amundsen Bay. *(P.G. Law)*

MV *Norsel* at Maudheim, 1950. *(P. G. Law)*

The Larsemann Hills *(P. G. Law)*

Four explorers, left to right: Phillip Law, Sir Douglas Mawson, General Hjalma Riiser-Larsen, Captain John King Davis, in Melbourne, about 1956.
(A. Campbell-Drury)

Ian McLeod (left) and Law at the Larsemann Hills.

Wyatt Earp at Hobart, December 1946-7.
(P. G. Law)

Pontoon landing, Macquarie Island. *(P.G. Law)*

Kista Dan held fast by pressure ice, 1954. *(P.G. Law)*

Heard Island Station . *(P. G. Law)*

Macquarie Island Station, 1961. *(P.G. Law)*

Salvaging equipment from the helicopter. The motor boat and the ship are in the background. *(P.G. Law)*

Hurricane easing — sailor tightening securing chains on DUKW. *(P.G. Law)*

14

MOUNT KOSCIUSKO THE SECOND TRIP

In 1938 I was teaching science at Melbourne Boys' High School. Bruce Osborne, a fellow teacher, was a fitness fanatic, a long-distance runner and a very enthusiastic skiing novice. At the school, we talked about skiing and about having some sort of ski holiday in the winter. As a result of my 1936 experience, I suggested that we should attempt to scale Mount Kosciusko via Hannels Spur from Geehi. This had never been done in winter and I believe it hasn't been done even to this day. We were young, optimistic, adventurous and rather foolish. I was just a novice skier and Bruce was even less practised. We knew very little about skiing techniques and we were fairly amateur campers, without any of today's sophisticated equipment. We decided, though, to attack Hannels Spur on skis in the August school holidays.

In mid-August we left Melbourne in my A-Model Ford and drove up through Albury and Corryong to Khancoban and Scammel's Waterfall Farm where we were to hire a packhorse.

The whole of the week before we left it had teemed raining in Melbourne and there had been continuous snow on the Alps. We set off with it still raining, thinking and hoping that it couldn't continue much longer and that we would then break into good weather.

When we reached Waterfall Farm, Scammel told us we would not be able to take the ordinary track into Geehi which I had previously walked—the one that had involved all the crossings of the Swampy Plain River—because the river would be flooded as a result of the heavy recent rains. So he took us on a hair-raising ride along the spurs and ridges.

We each rode a horse and Scammel led a packhorse with our gear on it. I was unaccustomed to horse-riding and the terrain frightened me, but these were mountain ponies, whose skills were quite remarkable. I was ignorant of this and to be traversing a very steep slope, with the ponies daintily stepping over fallen trees and branches and slipping and sliding on muddy gravel quite unnerved me.

We did not meet the Geehi Wall and I do not know to this day just what our route was that led us to the Geehi Hut. Scammel stayed overnight with us and set off back to Waterfall Farm next morning. We asked him to organise a search party for us if we did not return in a fortnight.

The next day is one which I shall never forget. We rose about 4 a.m., walked upstream from the hut to the two-wire suspension bridge, ferried our gear dangerously across, and then walked three miles up on the other side of the river to the point where Geoff and I had previously crossed back to reach Hannels Spur.

The river was running at full spate and looked very frightening. We stripped off and Bruce went first to essay a crossing. The water was thigh deep and at freezing point, because it was running straight off snow. He was halfway across when he lost his footing and was swept downstream. I rushed on a long detour around the fringing scrub of the river to reach the bank about 100 metres below and was able to wade in and grab Bruce's arm and the two of us struggled ashore.

Back we went to the crossing, determined not to let this obstacle block us. This time I stripped off and, taking one end of a climbing rope that we had, I half-waded, half-swam across to the other side. There I tied the end of the rope to a tree. Bruce secured the other end on his side and, in two or three trips, holding the rope, we were able to relay our stuff across. By then we were blue with cold and had to prance around on the bank to get warm, towelling each other down vigorously. We dressed and headed towards where the track began at the foot of the spur.

We had expected the snow line to be at about 300 metres but, to our dismay, we found it to be practically at river level. The torrential rain we had experienced in Melbourne had been extremely heavy snow in the Alps. The new snow engulfed the whole of the mountain, right down to the Geehi flat.

We were able to find the entrance to the Hannels Spur track, but it was a mess. First, the snow was half a metre deep, but second, the branches of the eucalypts that lined each side of the track had been weighed down by the weight of snow and had either broken off and become buried underneath the snowdrifts or else hung down heavily, completely obscuring the track. We found that if we beat the snow off these hanging branches with our ski-stocks, they would spring back to somewhere near their original positions and allow us to proceed. But as we thrashed our way up, with our feet deep in snow and our boots becoming entangled in the buried branches, we became more and more exhausted. Not only our footwear but our general clothing became saturated.

The following year, great bushfires swept through the Alps and burnt out much of the timber on Hannels Spur track. Not since then has the foliage presented the same composition and density as on our trip in 1938.

We reached a stage where we could not find the track ahead because the branches had fallen so that they completely obscured it. Then a strange thing occurred. We could see wombat claw-marks heading into the thick, impenetrable-looking scrub ahead of us. We found that, if we followed the wombat tracks and then beat the bush, the branches lifted and there was the track. This little wombat had apparently been used to running up and down the track when it was clear and now, even though it was obscured, he knew his way. By following these wombat tracks we were able to continue.

We became steadily more exhausted as, with heavy packs and skis over our shoulders, we plunged upwards through the snow and broken branches. Eventually we emerged from the trees and scrub onto the straight snow slopes at about 1800 metres altitude. It was a relief to take our skis off our shoulders and strap them to our feet. When carrying them they had kept getting entangled in the branches of the engulfing scrub.

We then ran into another difficulty. As novice skiers, we had never heard of 'skins' to attach to the bottoms of skis for climbing. We found that, as we headed uphill, our skis would slip backwards. 'Herring-boning' to prevent slipping is a very arduous and tiring exercise—it is much better to traverse a slope when climbing—but if you do not have waxes or climbing skins, your feet slip back each time you push them forward. So we improvised by cutting lengths of rope and winding them around our skis just behind the bindings and in this way we were able to proceed.

I have never in my experience of skiing in Australia seen powder snow so deep. In Antarctica, yes, but in Australia it is rarely cold enough. Each time we put our skis down we would sink up our knees in it.

As beginners, it was disconcerting not to be able to see our skis. Nevertheless, we struggled forwards with our roped skis sticking reasonably well. Eventually we were within striking distance of the top of the Townsend Ridge (named the Abbot Range, altitude 2000 metres) with only a couple of hundred more metres to climb, in that region of great boulders that I described in the earlier episode. Then the weather, which had been threatening, turned bad again and a blizzard developed with strong wind and heavy snow.

We realised that, if we crossed the top of the Abbot Range and went down into Wilkinsons Valley, we would probably get lost because visibility was zero. Also, after that, our chance of finding the Seaman Hut would be very small. By this time our drenched clothing had frozen solid so, each time we moved our legs, our trousers would crackle as we bent our knees. We were literally walking in sheath ice like armour.

Then we very sensibly decided to give up the project; but, before turning around, we sheltered behind a huge boulder, stripped all our clothing off and reclothed ourselves in fresh woollen underclothing right down to our socks. We then lit a small primus stove and heated up a brew of thick cocoa, drank it with biscuits and then headed back down the slope

We had taken the ropes off our skis so we could run smoothly but we had no idea how fast skis run on dry powder. Our method of descent was to proceed in a series of long traverses, crisscrossing our way down the slope, but at the end of each traverse we had to turn almost 180° to come back on the other traverse, and we were unable to do so in this deep snow. We had had no experience of turning in snow which was up to our knees deep. Our only way of dealing with the situation was to throw ourselves over at the end of each long traverse and then to get up and set off on the opposite traverse. At the end of that, we would throw ourselves over again. The problem was that the snow was so soft and deep that, when we threw ourselves over, we created a pit a metre deep. When we tried to rise by pushing our hands down, they just disappeared into the soft, fluffy material, so we wouldn't get any 'purchase' to push up.

Thus, at the end of each run, we would flounder around in the trough we had dug and it would take great effort and time, with our rucksacks on, to rise and set off again. In this way we descended 460 metres or so to the treeline, where we very gratefully took off our skis. Once again we were soaking wet.

From then on it was straightforward because the track had been cleared by us on the way up. We reached the river at about 1800 hours, having been on our feet since 0400. We had to strip and wade through the icy water carrying all our gear, dress again on the other side, walk down to the suspension wires, cross to the other bank, walk to the hut and there light a fire, cook dinner and fall into bed. Although it was a blizzard higher up, at the river it was rain. All next day it rained so we lay in the hut resting; but the rain did not look like stopping, so the following day we set off and walked, in rain, the 19 kilometres back to Waterfall Farm, including a number of river crossings. There we changed into pyjamas that we had in the car and set out on the long drive back to Melbourne. The book *Snowy Mountains Walks* (The Geehi Club) gives the distance from Geehi to Mount Kosciusko as seven miles and the climb as about 6000 feet. It advises walkers to do the walk downhill, not up, and states that a whole day should be allowed for the walk!

Several days later at school, Bruce Osborne said to me, 'I was just wondering, Phil, last night, what we would have done if one of us had twisted an ankle or broken a leg while on the mountain?' I replied 'You were thinking about it last night! I was thinking about it the whole time we were up there.'

15

AN ASTROFIX ON CLEFT ISLAND

In February 1966, 11 years after our journey to Lichen Island, I had one last chance to fill in some of the gaps in our knowledge of this interesting region, for I was to resign as Director of the Antarctic Division in April. The ship was *Nella Dan* (Captain Wenzel Gommesen) and on 15 February we had sailed from Mawson with Sandefjord Bay, at the bottom of Prydz Bay, as our objective.

The following day we moved eastward along the coast of MacRobertson Land and through the icebergs grounded on the Fram Bank north of Cape Darnley. The scenery was superb, with hundreds of icebergs—many beautiful, grandly-pinnacled glacier bergs, as well as the more common, prosaic, tabular type broken from the Amery Ice Shelf. There was plenty of space between them and we had no difficulty penetrating the accumulation. The echo sounder showed a bottom planed flat by icebergs over centuries with a depth of 120-130 metres.

We changed course so as to pass reasonably close to Cape Darnley, hoping to obtain a few sun shots to fix the position of the ship and then to use radar to obtain an accurate location for the cape. Its exact position had always been a little doubtful.

The captain was proceeding at slow speed despite the fact that the ocean floor was quite flat and he had more than 100 metres of water under the ship's keel. In my diary I noted: 'However, as I am not in a particular hurry, I shall not become impatient as I usually do in such circumstances, but merely register once more the unnecessary caution of the average captain.'

Cape Darnley appeared as a high dome of ice protruding from the Antarctic plateau at the north-west corner of Prydz Bay. The domed shape must have been due to a rock feature below, but there was no sign of rock in the ice cliffs rising from the sea. We rounded the cape between

1200 and 1400 hours and then ran into a mass of wedged bergs that completely blocked our path and forced us to deviate to the east. Between 1900 and 2000 hours we could see to the west the junction of the Antarctic plateau ice and the flat surface of the Amery Ice Shelf.

The usual katabatic wind was blowing down the slopes of the continental ice, streaming drift snow out over the vertical cliff faces across the sea. The first mate was doing a good job plotting the edge of the Amery Ice Shelf by radar and our speed was stepped up to nine knots in 240 metres of water.

I rose at 0730 hours on Thursday 17 February to find the ship doing 10 knots in deep water. However, another accumulation of icebergs forced us to move away from the ice shelf and the captain reduced speed to one and a half knots. This was infuriating. Our echo sounder recorded 870 metres, there were floating icebergs all around guaranteeing deep water, and the bottom was dead flat.

Finally we could see Mount Caroline Mikkelsen at the bottom of the bay so I went up the crow's-nest. From there I telephoned the captain, who was on the bridge, telling him what course to follow to approach the mountains, but he was reluctant to go. He said, 'There are a lot of icebergs down there.' I persuaded him to head for the first large berg, just to have a look, but he went to the wrong one and I had to ask him to proceed further to starboard. He inched his way along and finally stopped, looking at a floe that appeared to be grounded. I could see that I would be unable to get him to go further, so decided to push south by motor launch. However, it was then 1415 hours, with the best part of the day already gone.

At 1525 hours eight of us set off in the *MacPherson Robertson* launch—Edward, Oliver, Haigh, Cameron, Wilson, Corrie, Woinarski and I.

From the crow's-nest I had seen a low lip of ice protruding from the foot of Mount Caroline Mikkelsen and this was our objective. However, from sea level it was difficult to identify. When I saw a low ice edge with an ice slope beyond, we landed and roped up. After climbing across some crevassed snow and ice to the top of the slope, we found that we were on an iceberg! A novel experience! Our geophysicist, John Haigh, confirmed this when attempting to obtain geomagnetic readings—his instruments showed that we were afloat and moving gently. So we clambered down, re-embarked in the launch and proceeded around the iceberg towards Mount Mikkelsen.

Just before the mountain, we encountered a great bluff of rock rising straight from the sea so I decided to place Robin Oliver, our geologist, there. We landed to one side of the bluff and, as Haigh said that this location would be suitable for his measurements, we all went ashore except Edward, who stayed on board to tend the diesel engine, which was left running. Oliver collected rocks, Cameron lichens and Wilson a round of photos from the summit. Cameron also assisted Haigh with his magnetic observations. Corrie attempted to obtain an azimuth by sun observations, but was frustrated when the sun clouded over.

We found that we were on a small island just north at Mount Caroline Mikkelsen, giving a commanding view of all the features in the neighbourhood. We saw that we could not have landed directly on Mikkelsen, but would have had to walk half a kilometre over the fringing lip of ice I had noted from the crow's-nest. We had been fortunate to land where we did.

We stayed for two hours and departed at 2030. The weather was calm and enjoyable and the experience had been most interesting. Although I had flown over the area, this was my first view of it from the ground. To the south lay Caroline Mikkelsen and the Munro Kerr Mountains, with other nunataks fringing the eastern limit of the Lambert Glacier. On the western side of the glacier rose the continental plateau while, further north-west stretched the Amery Ice Shelf. To the north-east were cluttered bergs and the Polar Record Glacier Tongue. Adélie penguins were moulting on the island but we could not see any rookery. Mosses and lichens proliferated. The rock was granitic with charnokite and garnetiferous gneiss.

We chugged our way out past our iceberg and then around and between other bergs through brash ice. We had only a general idea of our direction, for ice blocked any distant view. We estimated that we proceeded 12-14 kilometres before reaching the ship.

Dinner was waiting for us and the captain immediately headed north towards open water. It was exasperating to see that he was now travelling at three-quarter speed whereas, on the way south, he had gone dead slow on the same track.

Friday dawned with an easterly wind of 20-30 knots. There was a light overcast and later it became misty with snow showers. When I went to the bridge after breakfast we were heading east on a course leading well north of Lichen Island, which I had visited in 1958. I was curious to see Lichen Island again so asked the captain to head for a large iceberg north of the island. He seemed reluctant to approach it and inched his way towards it at dead slow speed. I was so frustrated that I went down to my cabin and wrote for an hour. Again we had deep water (echo sounding 200-300 metres) and floating icebergs showed a lack of any shallow banks.

I gathered together Olive, Haigh, Cameron, Baggott, Bennett, Edward and Corrie and collected my cameras. When the captain reached the lee of the big berg at 1015 hours we departed in the launch with me at the wheel. The sea was rough and choppy. At Lichen Island we found a heavy surge running on the rocks and, although we might have put some men ashore, we could not have landed their instruments safely. Nevertheless, I drove around the island and refreshed my 1958 memory of it.

In direction 120° I saw another small island roughly six kilometres away and, beyond it, the Bolingen Islands closer to the continent. I decided to try for a landing on the small island in place of our attempt on Lichen Island. I was a bit concerned because there was a ring of icebergs around all except the northern approach to Lichen Island, and these would block the ship's view of us. There was also the captain's reluctance to take the ship south or south-east past Lichen Island. If, on our way to the other island, our engine were to fail and the weather were to deteriorate, we would be in trouble. I decided to risk it—it was my last chance as head of ANARE to make a new landing in this region that I had struggled to conquer for so many years.

We tossed and pitched our way across the intervening water and Corrie became seasick. Arriving at 1100 hours, we found that it was two islands, not one, with a narrow channel about six metres wide separating them. The northern island rose sheer from the channel and around the other sides, precluding any landing. The southern island offered the possibility of a landing if we could enter the channel; the surge on the rocks elsewhere was too heavy.

The easterly wind was funnelling through the channel, which was about 45 metres long, and driving a strong current through it. The question was, what was the depth of water in the channel? I would have to charge into the channel at speed to maintain steerage against the wind and current, for if I broached I would strike the rocks on either side. The steep walls of rock each side of the channel suggested that the cleft between the islands was deep. However, if we holed our launch here, the captain would have a serious problem trying to rescue us. I asked the members of the party were they prepared to risk it and they all said 'Give it a go.'

I took the launch in at three-quarter speed and was relieved to enter the channel without any crunching noises below. We went well in to where a narrow rock ledge a metre above the water on the southern island offered a landing place and there we bumped up and down roughly on another ledge of rock below the surface as, with engine and rudder, I held the launch with its starboard bow and side against the rock face. The heavy keel seemed to be taking the bumping all right and the hull was not scraping too badly along the side.

The men hurriedly disembarked and scrambled up to the summit of the island with their instruments, about 12 metres above sea level. I handed over the launch to Edward and went up to join the others. After taking various photographs I returned to the launch and, with Edward as passenger, took it out again to the open sea. Haigh had asked me to do this because the metal in the boat might affect his magnetic readings and I was reluctant to leave the boat bumping on the rock ledge for too long. Before leaving, I put ashore a survival kit and a tent together with some lemonade and chocolate. I took the launch out by continuing straight on through the channel against the wind; then Edward and I stooged around for an hour until the men ashore signalled that they had completed their work. I asked Edward to take the launch back into the channel while I took cine shots of our approach. I told him to take it fast.

However, Edward was over-cautious and approached too slowly, losing steerage way. He broached and drifted towards the cliff on the north side of the entrance to the channel, and the next moment, the boat was bashing and twisting on some submerged rocks below the cliff. Dropping my cameras I pushed him aside and grabbed the wheel as he was trying to back off. The next few minutes were hectic as we struck a reef astern, then I took the boat forward. But the wind swung us beam-on in the entrance and the bows crunched on the same reef. I was relieved to be able to back off and then, with some further manoeuvring forwards and backwards and a few more nasty crunching noises, to return to the open sea.

I lined up the launch on the entrance, gave it full speed ahead and stormed into the channel, putting the engine astern to take the way off as we approached the landing ledge. The men came aboard with their gear and we surged ahead and out the far side of the channel.

A cairn had been built on top of the island and under it had been placed a tin can containing a note of our landing and a small Australian flag.

We radioed the ship to say we had left at 1315 hours. I photographed Lichen Island as we passed but we did not investigate the possibility of another landing. The trip back was very rough, with a strong easterly wind on our beam, and my arms ached as I tried to steer and to hang on as we rolled and pitched. It was a relief to reach the ship, although we could not be hoisted until it had moved over once more into the calm water in the lee of the large iceberg.

I would dearly have loved to have crammed into our day a landing on the Bolingen Islands, but the weather was not good enough and time was against us. So this was one section of the Prydz Bay region upon which, over a period of 12 years, I had failed to land.

16

THE LAST VOYAGE OF HMAS *LABUAN*

In February 1951, HMAS Labuan *broke down returning from Heard Island and was lucky to reach Fremantle.*

When the ANARE was established in 1947, the major objective was the establishment of permanent stations at Heard Island and Macquarie Island. As Australia had no ship of a type suitable for voyages to these remote islands, the Royal Australian Navy was asked to help. After prolonged consideration the Navy offered to provide HMA LST 3501, a 'landing-ship-tanks' or LST. Such vessels had been designed by the USA for use during World War II to land tanks and other vehicles, with troops, on beaches during invasions. They were used in both the European and Pacific theatres of war.

An LST is an ungainly, ugly ship. The engines, the bridge and superstructure and most of the accommodation are aft. In front of these stretches a great, long foredeck from which rise 'Samson-posts' carrying derricks for lifting cargo. Beneath this foredeck is the 'tank space', a large hold for carrying the vehicles. The bows of the ship terminate in blunt fashion with two great doors that open outwards, allowing a ramp to be lowered. In use the LST runs its bows aground on the beach, the doors are opened, the ramp is lowered and the tanks drive out.

On each side of the tank space lie a number of bunkrooms to accommodate sailors and troops. Officers are quartered aft, beneath the bridge superstructure, where also are the dining quarters and other amenities.

When it was first proposed that an LST be used by ANARE, there was a great shaking of heads amongst knowledgeable mariners, who pointed to the numbers of LSTs that had broken in half in the Atlantic and elsewhere during the war. Admittedly, most of these were welded ships, built in a hurry under pressure from the demands of warfare logistics. LST 3501 was one of the few riveted ships and it was pointed out that, not only do such ships avoid the dangers of faulty welding, but the riveted plates permit some movement, and greater flexibility results.

In any case the decision was made to use LST 3501 and in the 1947-48 summer she made two long voyages—return trips of 11 200 and 2 900 kilometres to Heard Island and Macquarie Island. The misgivings that prevailed before these voyages were largely allayed as a result of these successful ventures, so the ship was sent south again in 1949, 1950 and 1951. In December 1948, she was renamed *Labuan*.

I happened to be leading the Heard Island relief expedition in 1951 on board *Labuan* when she finally cracked up, giving the pessimists the chance to say 'We told you so!' During this voyage the ship sustained severe bashing by successive storms to an extent that few ships have ever had to endure.

We sailed from Albany, WA, on Monday 22 January 1951, with Lieutenant-Commander Ian Cartwright, RAN, as captain. It was his first command. In the roaring forties we butted our way into the prevailing westerly winds. On 24 January, wind W Force 5, we ran into a heavy swell and big waves, and on 25 January the wind was NW Force 5.We were reduced to half-speed by heavy seas. On 26 January, wind W Force 5, our deck cargo of several hundred 44-gallon drums of fuel shifted and the ship was hove-to while sailors struggled on the sea-swept foredeck to tighten the restraining chains. On the night of 27 January, with the wind WNW Force 7 on our starboard quarter, the ship took a severe pounding. But worse was to follow.

On 28 and 29 January the wind exceeded Force 8 and the ship made little progress.

On the night of Monday 29 January I went onto the bridge to find hurricane-force winds and 13-metre waves. The long foredeck of the ship would run up onto and beyond the crest of a giant wave and then, as the wave passed amidships, the bow would come crashing down smack into the middle of the next wave. There would be a crash as of a thousand kerosene tins and the foredeck would oscillate like a springboard—wham wham-am, am-mmm—and a succession of kinky waves would move down along the deck for the length of the ship, almost pitching you off your feet as they passed.

The helmsman had difficulty in holding the ship's head into the wind. With its shallow draft and large area of bows, the ship would be caught by the gale and slewed off to one side. It would take five minutes of full rudder, with one propeller full ahead and the other full astern, to bring the head back into the wind, then a few minutes later it would veer off on the other side. We were all praying that the engines would not stop, for the LST could not have lived broadside in that sea!

Our men were quartered in one of a number of large bunkrooms situated alongside the large cargo compartment which comprised the main volume of the forepart of the LST. With them were a number of sailors.

In the middle of the night our men awoke to find no sailors in the bunkroom. They went and looked in the next, which was also empty. Then they panicked, thinking they had been deserted in an abandoned ship, so they dressed and rushed back aft. There, in the mess decks, fully clothed and with packed kitbags, they found most of the sailors, who apparently had reasoned that if the ship were going to break in two, they might as well be in the half that had the engines in it!

The night of 2 February was extremely uncomfortable, for the wind turned to NW Force 7 on our starboard beam and we were forced to change course to head into the heavy seas to avoid the excessive rolling.

On 3 February, the steering broke down in a Force 8 gale and sailors worked in shifts steering manually by the emergency gear. After half a day's work the steering was repaired, but, when changing over from manual to engine steering, the ship yawed off and, before she could be brought back into the wind, did three or four mighty rolls well over 40°, which is a lot for such a ship. A wave was shipped on the starboard side, which swept along the passageway on the main deck and poured down the companionway into the crew's mess decks. Our boys emptied out like ants in a couple of seconds flat, under the impression that the ship was sinking.

On Sunday 4 February the wind dropped to WSW Force 7, but there was still a six-metre swell and it was a great relief to all when, on Monday 5 at 0747 hours, we arrived at Heard Island. At last the wind had dropped and, in Atlas Cove, it was relatively calm. We were, however, faced with very heavy rain for our first day's activities, with 50 mm being recorded (the heaviest for 12 months). Then, when we went to open the bow doors to drive out our DUKWs, the ramp jammed and could not be lowered.

This meant that the heavy lift gear had to be rigged to the ship's derricks in order to hoist the DUKWs out through the hatches on the foredeck. The boatswain and his party had trouble assembling the gear and it was after noon before the lift could be essayed. By then a dangerous swell had risen.

Preparatory to hoisting the DUKW, a guy line had been attached to each side of it and sailors had been appointed to haul on these to steady the heavy vehicle. However, to my dismay, I noted from the bridge that, although several sailors were ready to haul on the guy rope on one side, the guy on the other side hung loosely down to the deck. When the boom was raised and the ship rolled to one side, the sailors pulled on their line and steadied the DUKW, but when the ship rolled to the other side there was no control to steady the DUKW, which swung away out over the side of the ship. The next phase of the roll sent the DUKW swinging back so hard that the restraining guy was torn from the hands of the sailors. The DUKW then swung violently from side to side as the ship rolled, with steel wire ropes lashing like snakes through the air and flying dangerously close to the hands and legs of the men on deck. The giant pendulum of a 7-ton DUKW swinging back and forth across the foredeck and bashing into 'samson posts' made quite a spectacle and it seemed certain that this valuable component of our landing equipment would be severely damaged. However, after several minutes of general panic on the deck, the synchronisation of the sea swell and the DUKW's oscillation was broken, the arc of swing decreased and the donkeyman at the winch was able to drop the DUKW down onto the deck.

Thereafter we managed to unload several DUKW-fulls of goods, but then the swell became too great and unloading ceased. Shortly afterwards, with the wind rising again, the *Labuan* was forced to weigh anchors and proceed to the open sea.

There followed the worst night of the voyage, with hurricane winds W Force 12 (80-100 knots) and the ship hove-to some kilometres off the island.

Next morning at 0815 the steering failed again in heavy seas, but fortunately was fixed within a few minutes. Later, the *Labuan* entered Corinthian Bay with a view to anchoring but had to put to sea again.

On Wednesday 7 February the ship entered Atlas Roads at 0450 hours but found conditions unsuitable, so went around to Corinthian Bay, where it was able to anchor. Unloading commenced, but was very slow owing to various delays with DUKWs, mainly at the ship's side. Then I received a message stating that the engineers were having trouble with the ship's boilers. The bashing the ship had taken had shaken the condenser tubes immersed in the cooling salt water tanks. The rivets had loosened and allowed salt water to enter the tubes. Mixed with the fresh condensed water, the result was that, when the water returned to the boilertubes, salt caked out and clogged the boilers, partly insulating them and reducing their efficiency.

Another problem arose. The captain notified me that sea water was entering some of the fresh water tanks and that water shortage would force him to leave Heard Island that night to proceed to Iles Kerguelen, 290 kilometres NW. There he would replenish the water supply and make repairs to the condensers. Then, if weather permitted and the ship were in a seaworthy condition, he would return for a short visit to unload the remaining stores.

This was a very serious situation. The continuous foul weather, the mechanical faults on the ship and the physical stresses involved for all concerned had reduced morale on board *Labuan*, particularly amongst the crew, to rock bottom. No-one trusted the ship not to break down and the prevailing opinion was that, once we had left Heard Island, it was most unlikely that we would return.

I was forced to work on the assumption that there was a possibility that we might not return. For the wintering party to survive there were certain essentials, particularly fuel, that still had to be unloaded. We made lists of the minimum requirements and worked furiously through the day and evening to get them ashore.

An unexpected crisis then confronted me. Dr Rec, the medical officer for the wintering party, had his foot jammed between a DUKW and the ship and had two bones broken. The question was, should I leave him at Heard Island? If not, what would the party do for a medical officer? Would it be fair to ask Dr Gwynn, the outgoing MO, to stay on for a second year?

Later that night the swell arose again. The result of all this was:

- It was decided Dr Rec could remain;
- Three DUKWs were left ashore because the swell was too bad to permit their being taken on board;
- The coke and range fuel needed for cooking and heating had not been unloaded.

It was a very worried PGL who sailed that night at 0103 hours for Kerguelen! And it was flat calm all the way. Unbelievable!

We stayed at Kerguelen, where the French had an expeditionary establishment, for three and a half days. The weather was kind, the sun shone, the French were extremely hospitable, we loaded our fresh water and the engineers repaired the condensers.

Morale is a strange thing. Now, at the end of a good three days, no-one seemed to have any doubts whatever about returning to Heard Island!

We anchored again at Corinthian Bay on Tuesday 13 February at 1547 hours. There was some wind, but it was offshore and there was no swell. Later, things improved even further. We worked furiously unloading our remaining cargo and I was surprised at how much there was! It would have been a poor year for the wintering party if we had not returned! By midnight it was all ashore.

Next day, Wednesday 14th, we built huts and did other jobs. On the Thursday the ship put to sea again in bad weather. We were to have departed but it was not possible. The Friday was even worse and we ultimately got away at 0832 hours on the Saturday.

Diary entries I made during the voyage back to Australia tell the rest of the story of the breakdown of *Labuan*:

Sunday 18 February

Wind WSW Force 2.

The steering engine aft has 'had it' apparently. The main shaft of the electric motor is so worn that the bearings become chewed up just as quickly as they are replaced. Yesterday crew members were rostered to steer manually, but it was most laborious, with two men both on the wheel for half-hour shifts.

Then an engine-room artificer had a brilliant idea. He realised that there was a large electric air-circulating fan on the deck above the steering engine, so he drilled two holes in the deck and ran a belt from the fan engine down to the steering mechanism. This worked, and it continued to work for the rest of the voyage. Upon returning to Australia, the artificer had to face a court-martial for drilling holes in the ship's structure without approval from the Naval Board!

The captain decided to head for Fremantle, not Melbourne.

Monday 19 February
Wind WSW Force 2.
The ship's engines stopped for an hour in the afternoon. Luckily it is fairly calm, with a swell of only about three metres.

Tuesday 20 February
Wind SW Force 4
In the afternoon the engines were stopped for eight hours while a burst boiler tube was removed. The poor chief engineer has had a rotten run—he is just about worn out. Everything that could happen, just about, has happened.

We are going to Fremantle for repairs to the steering engine and to have the boilers cleaned. It is said that this will take a week. Then they expect further work on the ship in Melbourne will need a month. The poor expeditioners at Macquarie Island will be getting very impatient waiting for us to relieve them!

Wednesday 21 February
Wind E Force 1.
Today the main boiler-room fan broke down and we remained stationary for about eight hours while repairs were made.

Thursday 22 February
Wind NE Force 3, then N then NW.
When the wind turned NW we rolled badly. Water rationing has been imposed. Drinking water only, and that only at meal times. At other times one must drink beer (of which there is an ample supply).

Friday 23 February — Calm

Saturday 24 February — Calm

Sunday 25 February — Calm

Monday 26 February
Our bad run continues. The large refrigerator in the hold which we have used to store biological specimens has broken down and our ornithologist, Norman Favaloro, has lost all his birds!

Tuesday 27 February — Calm

At 0300 hours the engines stopped—our remaining fuel is contaminated with sea water and two more pipes in the boiler have burst. At 0700 hours we are rolling idly, wailing for a tug to steam out 200 kilometres from Fremantle to take us in tow.

I suppose we are lucky to have got this far! Fortunately it is still calm.

We are now derelict. No engines, no electricity, no pumps, no lights, no water (either fresh or salt), no flushing of toilets.

The tremendous punishment the ship has taken on this voyage, added to that on the tough trips in 1948, 1949 and 1950, has finally finished off this ship. The riveted plates have loosened and allowed sea water to enter the fuel tanks and the water tanks. The condenser tubes, as previously explained, have shaken loose their bolts, and cooling salt water has entered the tubes. The salt from this has clogged the boiler tubes, which have burst.

A crack has opened up across the foredeck, running directly through the steel plates (not along the rivet lines) just in front of the 'samson-posts'. If the ship were to break in half, this is how it would start.

The tug arrived at 2200 hours but decided, as it was fine weather, not to hook up until daylight tomorrow.

Wednesday 28 February — Fine, calm

At about 0200 hours the engines started up so we proceeded without using the tug at about five knots.

However, at 0600, the engines stopped again. The tug then took us in tow. We arrived in Fremantle after dark and were not permitted to berth. Tins of fresh water totalling 100 gallons were sent out to us for drinking.

Thursday 1 March

We berthed at 0900 hours.

How lucky we have been. It is quite remarkable that, after weeks of storm, we should have had six days of calm weather to coincide with the ship's deterioration. One really heavy blow over the last few days would have sunk us.

17

DUKW WRECK AT MACQUARIE ISLAND

Macquarie Island, 54°30'S latitude, 159°E longitude, is a fascinating place. Swept by the westerly gales of the Southern Ocean, its coast is as stormy as any in the world. The weather is tempestuous, cloudy, wet and cold, with rain or snow falling on roughly 330 days a year. The temperatures range is from 20°F to 50°F, the average over a year being around 40°F.

The island is roughly 34 kilometres long and 5 kilometres wide, with its long axis running NNE-SSW. Behind narrow, pebbly beaches or rocky strips a steep escarpment rises to a plateau, from which a few higher hills reach a maximum altitude of 440 metres. On most days, the plateau is obscured by cloud. There is no permanent ice or snow and there are no trees or bushes, the vegetation consisting of a variety of grasses, mosses and lichens, of which a coarse tussock grass is the most common. The island provides nesting grounds for numerous flying seabirds and several million penguins, while, in the spring, some 100 000 elephant seals haul out to breed.

Macquarie Island is a dangerous place to approach. A number of sailing ships in the nineteenth century, on the 'great circle' route to South America, came to grief on her reef-girt western shores. In such a southern latitude a ship might sail for days without sighting sun or stars for an astrofix, so its captain would be uncertain of his position within 50 to 100 kilometres. Driving in an easterly direction, in poor visibility with a strong westerly wind astern, the vessel would face a considerable risk of piling up on the reefs of this shore. I remember visiting the island in 1952 in the small Norwegian ship *Tottan* with an unreliable radar. To play safe, the captain deliberately passed 100 kilometres south of the assumed position of Macquarie, then beat back to approach it from the east, against the wind.

The ANARE station at Macquarie Island was established near the northern extremity of the island in March 1948. It has been maintained continuously ever since.

Some of the most dangerous episodes of my career have occurred during landings to resupply the station or to establish and resupply field huts on both the eastern and western sides of the island. Heavy surf, great floating fields of giant kelp, and reefs and 'nigger-heads' of exposed rock fringe the shores, while gale-force winds, fog and snow showers add to the navigational problems.

Over the years, the men of ANARE have had numerous dangerous and exciting experiences loading and unloading ships, of which the following are a few typical examples.

When the *Labuan* broke down on the return voyage from Heard Island early in 1951, we were left without a ship to relieve the Macquarie Island station. We found it extremely difficult to find a suitable ship in Australia, one with enough endurance to travel 4200 kilometres on the return trip to Macquarie without refuelling. There was no time to seek a charter from overseas if we were to carry out the relief before the winter.

Our shipping agents finally found a vessel, the *River Fitzroy*, a cargo ship built to transport iron ore. She had very limited passenger accommodation and we had to convert storage space in her forecastle into sleeping quarters for our expeditioners. The steel hull was not insulated or lined and, in the cold sub-antarctic seas, condensation dripped continuously from the decks and bulkheads. To reach the general accommodation amidships—dining saloon, lounge room, etc.—the men had to traverse the foredeck, which, in bad weather, was swept by spray and shipped seawater. Safety lines were rigged to assist them essaying this risky passage.

The *River Fitzroy* had a displacement of about 8000 tonnes, much too big and clumsy for work on a rocky shore with no harbour facilities. She was crewed by sailors of the coastal merchant marine, none of whom had had experience of the sorts of operations ANARE normally carried out or of the conditions involved. During the voyage we had problems of men pilfering the ANARE cargo and, at Macquarie Island, problems with union decisions resulting in the crew refusing to carry out deck duties under conditions regarded as normal by crews of our previous expedition ships.

Our problems began in Melbourne, when, at 1930 hours on

Wednesday 25 April 1951 the *River Fitzroy* arrived in Melbourne from Whyalla. Although late in the day, I sent men on board to make a preliminary survey of the alterations to the ship that would have to be made. Haste was needed, for the ship was due to sail in three days' time.

Motor transports had been loaded with cargo at our Tottenham depot over the preceding days and on the Thursday began arriving at the wharf. But it rained most of the day and the stevedores would not unload them. The only breaks in the weather occurred during 'smoke oh' rest breaks or at lunchtime!

In order to accommodate the sheep we wished to send to Macquarie, Works and Housing Department carpenters constructed sheep pens on the deck, but a stock transport inspector visiting the ship ruled that we could load only 10 out of our total of 22 sheep. I had arguments with the ship's cook, pantryman and chief steward because I had put on the sheep pens before loading the ship's meat; but the truck carrying the meat stood idly beside the ship waiting for the rain to stop, then finally had to be sent back to wait overnight at the wholesale butchers.

Next day—Friday 27 April—the ship was to proceed at 0600 hours from No. 14 wharf to No. 16 to gain access to a dock crane which was to load a bulldozer, the rubber pontoons and the Army DUKWs. However, owing to crew trouble, a ship that was there could not move out, so our move was postponed 24 hours. The motor transports carrying the bulldozer and the pontoons had to be sent back to Tottenham. All very frustrating. Finally, government navigation officials ruled that we could not carry two passengers in the ship's hospital, which was part of our plan to overcome the accommodation problem. We therefore had the chief steward lock up the hospital (but he opened it again and we used it after the ship had finally sailed from Melbourne).

Difficulties continued to arise. The chief steward and I had separately telephoned the captain, Monty Mathers, who had agreed verbally to each of us that mattresses for the passengers were to be supplied from the ship's stores. But when the time came to install them, the captain denied this, insisting that I had agreed to provide them. Eventually, after considerable confusion all around, the ship supplied them.

Army DUKWs, stripped down, weigh 6.5 tons. The ship's derricks were certified to take loads of five tons. The captain was prepared to take the risk but the first mate would not agree, so I had to arrange for heavy 'jumbo' gear to be provided for use in hoisting the DUKWs.

We were accustomed to carrying the DUKWs on top of the hatches on the main deck, but the first mate ruled that they must be placed in the hold. When Army Captain Eric Troy arrived at the ship's side with his DUKWs at 0800 hours on the Friday, the mate told him to take them away because he could not load them onto the deck until all the cargo was inside the hold and the hatches were closed. Either he had changed his mind or the captain had overruled him. Such problems had been exacerbated by the fact that the captain had left the ship and gone home for a break. We had no means of contacting him.

At 1700 hours on the Friday, the chief steward was very drunk. When the ship's meat (for the second time!) arrived by truck at the wharf, he said it was our meat, not the ship's. Further, he would not allow us to load ours until his had been put into the ship's refrigerator. After the meat truck had stood in the cargo shed for most of the day I ordered the stevedores to load it onto the ship and place it in the 'tween-decks space in the hold. When this had been done we found that 80 per cent of it was ship's meat!

We sailed from Melbourne for Macquarie Island at 1230 hours on Saturday 28 April. It was a rough trip. With the ship on a course SSE, the prevailing westerly winds were creating heavy seas that caught the ship on the starboard beam and caused it to roll heavily.

On Friday 4 May there was a wind of Force 8 and waves up to 11 metres in height. A number of seas came over the side and crashed with great force against the bulkhead of my cabin on the main deck. One sea crashed onto the deck above and ran along to flood the port passengers and the officers' cabins, past the hospital (whose door was fortunately closed) and down into the engine room. Another sea, we found later, had staved in the cabin shelter and hull of our motor launch on the main deck. It was a relief to arrive at Macquarie and drop anchor in Buckles Bay on the eastern side in the lee of the island.

The hazards of landing operations at Macquarie Island are illustrated by the following extracts from my voyage diary:

Friday 11 May. Macquarie Island

A cold front had passed through during the night and we awoke to a westerly gale of 50 knots with driving rain. When the rain stopped, Captain Troy (on board) asked me (ashore) to send out a DUKW to try to unload drums of fuel. The DUKW went out all right, with the wind astern, but two DUKW hands were nearly swept off the front of the vehicle when attempting, in a heavy swell, to hook the DUKW's bow line to the mooring line stretched down the side of the ship. Giving up, the DUKW turned, empty, to return to the beach, heading straight into the gale which, blowing off-shore, had whipped the water into a sea of white horses.

The spectacle of the DUKW bashing its way towards the shore was magnificent! Once each second she struck a choppy wave, sending a cascade of water 9 metres into the air which the wind whipped back as a white wall of spray for 30 metres astern. One man, geophysicist Hugh Oldham, stood in the DUKW, head and shoulders above the windscreen and with sluices of icy water whipping his face, to direct the driver, who could not see a thing. From where I stood on the shore the DUKW was only occasionally visible; all one could see were the repeated bursts of white spray through which, intermittently, could be observed the dark blur of the DUKW's hull and the darker vertical blob of the navigator.

Friday afternoon

At about 1500 hours the two rubber pontoons, used as unloading barges, that had been lashed together and anchored just off the landing beach, broke loose and began to drift rapidly out to sea towards South America. My deputy, Dick Thompson, gave the alarm, and, within minutes, there was a crew of Thompson, Jacka, Bunt and McCormack in one DUKW giving chase and another DUKW standing by on the beach, with Flanagan and Wyatt, ready for a rescue if necessary. The gale by now was at its height, with gusts over 60 knots. There was no swell here on the lee side of the island but the fierce wind created steep, choppy waves whose crests were whipped off and flung as spray and spume over the surface of the sea.

Again the DUKW had no trouble going out and caught the pontoons about a mile offshore. They managed to put Thompson and Bunt onto the pontoons and secured them with a heavy line to the DUKW; then they turned back towards the shore. From then on, until they reached the shore, we hardly saw them—just a blur of white for the DUKW and the occasional glimpse of two black blobs behind, which represented the two men on the pontoons with spray and green seas sweeping over them.

Saturday 12 May

At 0645 the first mate said he was ready to try unloading the D4 Caterpillar tractor during the morning for, although there was a strong NW wind, here on the lee side of the island there was little swell. One of our DUKWs was a bit sick—probably water in the fuel—but at 0930 hours I sent out the other one, hauling three rubber pontoons lashed together to act as a raft for the bulldozer. Heavy timber bearers had been laid across the pontoons to act as decking. Two days earlier, the captain had asked could we dismantle the tractor and unload it, part by part, using a DUKW for transport instead of sending the complete vehicle ashore by pontoon raft. He was afraid that the whole tractor, swinging on the derrick in the swell, might punch a hole in the side of his ship. (That possibility certainly existed!) However, I had to refuse; it was some years since the tractor had been dismantled and there was no guarantee it would come apart without great difficulty. And, in any case, I could not spare an engineer from the station for two days to do the job. I said that if the swell did not decrease to a point where he would be willing to unload the tractor, the station would have to do without it. Next day, however, he told me he had had some ideas about unloading it and was willing to give it a try. So here we were, ready to go. I should mention that we had removed the bulldozer blade to lighten the load on the derrick and had arranged for it to go ashore on the pontoon beside the tractor.

It took a considerable time for the crew to lift out the tractor onto the pontoon raft and to follow it with the blade. Then the DUKW towed the raft slowly towards the shore through the very choppy sea. The tractor was lucky to reach the shore because, under the heavy load, one pontoon developed a leak and the whole load slumped down on that side. It looked as though the tractor would capsize the raft. However, it made the distance, but then the fun started, for there was quite a surf running.

The first essential was to drag the array into a position where the three long pontoons, lashed side by side, lay with their long axis at right angles to the beach. While the beached DUKW kept the tension on the tow line, men from the pontoons plunged into the surf to haul guy ropes ashore, while men from the shore waded waist-deep into the surf to attach extra guy ropes to the pontoons. With a host of helpers hauling on the lines and battling with surf, the pontoons sluggishly swung into the right position. A cable was hurriedly connected to the bulldozer blade and linked to the DUKW's winch. Within a few minutes this heavy, clumsy article had been winched ashore. Other men on shore then dragged 'Bailey bridge' sections to connect the forward edges of the pontoon decking to the beach and the tractor driver quickly humped the vehicle clumsily across these onto the shingle.

The tractor was barely ashore when the pontoon array slewed around broadside to the surf and the bridging fell away. We had been lucky.

We next hauled the Bailey bridge sections up onto the beach and tried to do the same with the pontoons, but the group of three proved too heavy, so we had to disconnect them. This involved four of us kneeling on the edges of the pontoon decking, using spanners and hacksaws to unscrew or cut through the heavy bolts that held the deck timbers, and thus the pontoons, together, while the raft bucked and twisted and heavy waves broke completely over us. The separated pontoons were then successfully dragged up onto the beach by the tractor. By then it was lunchtime and we were all very cold, tired and hungry.

Sunday 13 May

We had two jobs left to complete our relief task—to put stores ashore at Sandy Bay and at Lusitania Bay, at each location there being a field hut for use by biologists.

On this Sunday I rose at 0545 hours and found the plateau covered with snow and the wind blowing bitterly cold from the SW. This presaged a fine day. I had planned for the ship to sail south to Sandy Bay, about eight kilometres, and for the two DUKWs to cruise down under their own power to meet us there. In this way we would avoid delays resulting from having to load and unload the DUKWs from the ship.

However, there was the usual frustration. We arrived at Sandy Bay about 0900 hours and had to wait until 1030 hours until the DUKWs turned up. The 'sick' one had caused trouble again, delaying their departure from the station and, just as it reached us, it broke down again. So we loaded up the good DUKW with cased stores and went ashore, leaving the other one moored astern while its driver replaced a carburettor.

Sandy Bay is a pretty spot. Despite reefs, niggerheads and kelp, there is a good approach to a fine sandy beach, littered with basking elephant seals and groups of Gentoo penguins. Behind this, a fine green grassy slope runs back to where the plateau escarpment rises steeply to the snow-clad plateau. The sun shone and the sky for a while was clear and blue. We unloaded quickly, carried the stores into the hut nestling above the beach on the grassy slope, picked up some cases of biological specimens, then hastened back to the ship for lunch.

As the next leg of our trip was longer—25 kilometres—and one DUKW was unreliable, we decided to tow the DUKWs down to Lusitania Bay. But it took the crew an age to rig two lines for the DUKWs and then they snapped as we started to move and had to be replaced.

We arrived at Lusitania Bay at about 1500 hours with the sick DUKW taking in water so rapidly that the use of the mechanical pumps plus one man baling were necessary. I gave orders for it to be hauled up on deck, while we loaded stores into the other DUKW and cast off.

On board were, in addition to myself, the following men: Captain Troy (driver), Thompson, Jacka, McCormack, Lindholm, Bunt, Wyatt, Ward, Taylor and Shipp. The ship was 800 metres offshore and it did not take long before we were cruising outside the surf line and looking for a place to land.

Normally in such an operation we would have some boat standing by in case of an emergency, but our only other DUKW was now unserviceable and the motor cutter we had put on the ship in Melbourne had been put out of action when struck on deck by a wave that crashed over the ship's side.

We wanted to land reasonably close to the hut, so that the distance to carry the cases would not prove too exhausting. There were two obvious approaches, A and B. It was difficult to decide as we rode up and down on the waves 100 metres offshore, but I thought A looked the more promising. (Later an inspection from the beach proved this to be the right decision, as the water at B was littered with submerged rocks.)

I asked Troy to approach A more closely so as to have a better look, but a large swell picked us up and carried us shoreward. Troy barely had time to straighten out the DUKW and point it towards the beach before we were creaming in on a huge breaker, like a surf boat at Bondi Beach. Ahead were clumps of kelp, several niggerheads and a beach of shingle and rocks. Troy steered the DUKW so far as he was able without broaching and dodged the visible obstacles; then, just as we were congratulating ourselves upon having come through successfully, there was a heavy crunch beneath us and, 10 metres out from the beach, the DUKW swivelled around, pivoting amidships on a submerged rock. Racing the engine and swinging the steering wheel, timed to coincide with the surges of succeeding waves, Troy was able to bounce the DUKW forward with several more grinding crunches, over some other rocks, but the heavy surf swung the DUKW broadside during the process and we finally found ourselves parallel to the beach, with the shoreward wheels grinding in shingle and those to seaward turning freely in deeper water. We were making no progress, the grounded wheels were becoming bogged and successive waves were breaking over us and filling the DUWK cargo well with water.

Troy ordered us to unload the cargo to lighten his vehicle, so we jumped out into the waist-deep, bitterly cold water and struggled back and forth between dry land and the side of the DUKW, carrying heavy cases of tinned and bottled provisions, stumbling as our feet sank into the shingle, falling and picking ourselves up, and all the time aching with the intense cold of our immersion.

One case escaped clutching hands and began to be swept out by the backwash, so Bunt plunged after it and wrapped his arms around the floating box. He was dragged off his feet by the backwash and I thought we would lose him as he disappeared seaward, but the next breaker picked up the floating case and dumped it and Bunt high up on the beach.

When the DUKW had been emptied, Troy directed us in a number of manoeuvres aimed at freeing the DUKW. He first tried rocking the vehicle, forward and back, forward and back, using the gears, with us all pushing in the surf, and had some early success; but then the wheels dug into the loose shingle and bogged down to the axles.

He then ran out a cable from the DUKW's winch and fastened it around a large boulder on shore. Using engine power on the wheels and propeller, and winch power on the cable, he tried to move the DUKW, but the overload pin snapped and the winch cable went slack. (The overload pin is designed to snap in case of an excessive load on the cable, to prevent damage to the winch. The pin is tested to withstand a certain force.)

Troy swore and set the motor-driven bilge pumps working to empty the half-filled DUKW while McCormack fitted another pin. Just as it seemed that the winching might yet get us out, the second pin sheared off. And then the clutch, which had been bearing the brunt of all the action—wheels, propeller and winch—gave up the struggle. The wheels ceased to turn, the propeller and the bilge pumps stopped and the DUKW rapidly filled with water as the surf crashed over it. Troy just had time to unfasten the battery and carry it ashore before the whole engine compartment was flooded. We stumbled up the beach quite exhausted and staggered to the hut.

By this time, it was quite dark and light snow had begun to fall. We saw the lights of the ship disappear as she headed out to sea. We were soaked to the skin and half-frozen.

The hut was just a makeshift job, being an old aircraft crate landed from the *Labuan* a couple of years earlier. Its rough timber planking had been caulked with tarred sisal paper. It measured 3m x 3m x 2.75m. While some of us cleared out the hut, the rest broke open wooden packing cases and built a large bonfire just outside the door. Inside the hut we lit lamps and a kerosene stove. We roasted ourselves around the fire until we were relatively dry, then cooked a meal out of tins—tomato soup, beef stew and plum pudding. Someone produced a couple of bottles of whisky and the contents rapidly disappeared. At about 2300 hours we bedded down, 11 men, on wood-wool, brown paper and other packing materials from the cases. In the 3m x 3m space, two swung in hammocks, one man slept on the narrow bench table and one underneath it, while the other seven lay overlapping on the open floor. We looked like sea elephants in a beach wallow! Luckily we had plenty of blankets.

While dinner was being prepared, Ward and Jacka had unpacked a 'sledge radio' (portable transmitter and receiver) which had been included amongst the supplies we had brought for the hut. With this, Ward was able to contact the ship and Macquarie Station and inform them of our situation. He arranged for a scheduled radio contact (a 'sked') next morning at 0700 hours.

It was clear to me that the DUKW was a total loss. We could not spare the time to bring block and tackle from the ship to haul it out and, even if we did, there was no way we could return the DUKW to the ship. Left for a year at Sandy Bay, it would not be worth salvaging.

What was of more concern to me was how we were to get back to the ship. The other DUKW was unserviceable, the motor boat had been bashed in, and the ship's lifeboats were too large and clumsy. Whatever we used, it would be a risky venture on this surf-lashed, rock-strewn coast. But I was too tired to worry much and left all planning until the morning.

Thompson, who was the man under the bench (and in darkness, away from the lamp) had performed prodigiously, in his usual energetic way, during the whole episode and then, after a few whiskies, was the first to fall asleep. Just as the rest of us were beginning to doze off, he woke suddenly, sat up, and cried out in alarm, 'I can't see where I bloody am!' And after a moment, 'Am I safe? Am I safe?

Troy, quick as a flash, grabbed an armful of woodwool, thrust it into Thompson's arms and said, 'Hang onto this, mate, you'll be all right.' Thompson clutched the bunch to his chest, lay back and fell fast asleep immediately.

We rose at 0600 next morning to a scene blanketed with newly-fallen snow. It was cold emerging from our small warm hut. I had a radio sked with the ship at 0700. In Melbourne we had put aboard *River Fitzroy* an eight-metre whaleboat with oars, as an emergency craft. I asked the captain to prepare it for use and arranged a further sked at 0800.

Then, with a couple of others, I looked for a landing place. We first examined the beach near the penguin rookery. The approach was cluttered with giant kelp and dotted with rocks and reefs—hopeless. The beach near the hut would have been suitable, but the only clear approach was now blocked by the stranded DUKW. Meanwhile, Shipp had walked further south along the shore and had found a narrow approach, reasonably protected from the swell and without any visible rocks.

The 11 of us gathered together and discussed the situation. We decided upon Shipp's beach and then worked out a system of signals to use in directing the whaleboat's approach, as follows:

• Orange flag (two 'Mae West' life jackets strapped to a pole)	—	To indicate the landing beach
• Two flags held horizontally, one each side	—	Come in
• Two flags held vertically	—	Stay where you are
• Two flags vertically and waving	—	Go back out to sea

We transmitted this information to the ship on the 0800 hours radio sked, then had breakfast. After this we carried the signal flags a quarter of a mile south to Shipp's landing spot.

The ship steamed in towards the coast well to the north of our position. For such a big ship it came in frighteningly close to the coast to launch the whaler. It had advised us that the whaler would be manned by Second Officer Bennett, at the rudder, with Flower and three sailors manning the oars. The boat was undermanned for these conditions, but if more oarsmen had been added there would not have been room for the eight of us who were to be taken off. (The three members of the 1951 Macquarie party who were with us would stay on at the hut.)

The wind was blowing from the NNE and Captain Mathers had shrewdly launched the boat a long way north of us, so the rowers could use the wind and swell to approach on a long, slanting run to our beach. After sending the boat on its way, the ship had turned to port and steamed out to sea.

It had begun to snow again and we couldn't see the ship. The next minutes were very tense. If we wrecked the whaler there was no other available craft, and there was always the danger of men being swept away to drown.

The whaleboat came in almost parallel with the shore, past the rocky point on our left, and, when it was opposite us and just outside the surf line, we signalled it with two vertical flags. Bennett dropped a bow anchor and waited a few minutes while I examined the swells and the breakers. When it appeared that the heaviest waves had washed in and there might be a bit of a lull, we signalled 'Come in'.

I expected them to turn the boat and back in, controlling their speed by hauling on the anchor line; but, instead, they rowed in bows foremost. We lunged into the surf, up to our waists, and grabbed the gunwales to hold the boat from drifting onto rocks that lay only three metres or so on either side. Then, with a sailor in the bows hauling on the anchor line, the boat came around beautifully and we just had time to haul ourselves over the sides and grab an oar each when the next run of breakers began. Rowing hard, with the bow man hauling on the anchor line, we bashed our way out through the breakers into the relatively smooth water beyond. Retrieving the anchor, we looked back and waved goodbye to Bunt, Lindholm and Wyatt who, a couple of days later, would walk back overland to the station.

The whaler was half-full of water and we were all soaked. It continued to snow and we rowed hard to try to get warm, while two men baled out the water. It was about three kilometres out to the ship and, despite our exertions, we became colder and colder. With aching arms and numbed fingers, we were only just able to climb the rope ladder hanging from the rolling ship on our arrival. The crew hoisted the whaler on board and the *River Fitzroy* set course immediately for Australia.

18

ESTABLISHING DAVIS STATION

In January 1957, ANARE established Australia's second Antarctic station, Davis, in time for the commencement of the International Geophysical Year (1957-58). The establishment of Davis Station was one of the major accomplishments of the ANARE, but it was a difficult exercise which, at one stage, was almost abandoned.

Before I embarked upon my Antarctic career, I had been intrigued to read that the US Antarctic Expedition 'Operation Highjump' had discovered, from the air, in February 1947, an extensive 'oasis' of ice-free rock, studded with lakes, at longitude around 100°E, that it had named the Bunger Hills. This made front-page news and I had followed with interest the discussions that arose concerning various theories explaining the origin of this unusual feature.

After I joined ANARE and began investigating what was known about Australian Antarctic Territory, I found that a similar 'oasis' had been discovered by the Norwegian, Captain Klarius Mikkelsen, on 20 February 1935. He had landed with his wife and seven men in a small bay on the most northerly part of this ice-free coast and had named the region the Vestfold Hills because of its resemblance to the province of Vestfold, south of Oslo in Norway.

It was natural, therefore, that I should consider the possibility of exploring this region when I was planning the 1954 ANARE voyage that was to establish Mawson Station. In due course, on 3 March 1954, I landed with several of my men from *Kista Dan* on Magnetic Island and on the mainland nearby. I followed up on this with a more extensive investigation of the area on 30-31 January 1955. I was greatly impressed with the potential of the Vestfold Hills as an area for scientific investigation and as a location for the establishment of a future ANARE station.

In 1956, 11 nations were preparing for Antarctic programs under the aegis of the International Geophysical Year. It was obvious that, if Australia did not establish a base in the Vestfold Hills, some other nation from these 11 would. I therefore requested, and obtained, approval from the Australian government to build a station there, if possible, early in 1957. During 1956, the huts were designed and prefabricated, the generators, radio and other equipment were purchased, and a team of five men led by Robert Dingle was selected for what was to be a small establishment wintering party. The ship *Kista Dan* was chartered, the master of which was Captain Kaj Hindberg.

The itinerary I had planned for 1957 involved relieving Mawson Station first and then proceeding to the Vestfold Hills. However, this was not to be. At 0220 hours on 4 January 1957, the RAAF pilots, Clemence and Grove, took off from a pool in the pack ice at position 66°10'S latitude, 65°21'E longitude, on a reconnaissance flight to Mawson. Returning at 0500 hours they described the ice as, first, heavy pack and then, fast ice for the last 64 kilometres. The chances of pushing through to Mawson looked remote, so I decided to have a look at the Vestfold Hills.

On 5 January I arose at 0400 hours, had a hurried breakfast and boarded our Beaver aircraft with pilot Peter Clemence. The return trip to the Vestfold Hills took roughly seven hours and covered almost 112 kilometres. We had some concern after passing Cape Darnley, for we had to cross Prydz Bay, which was mainly open water with a 40-knot wind whipping the sea into white-crested waves. Over the centre it would have been half an hour, either way, to reach safe landing areas in pools sheltered by pack ice.

During the reconnaissance I made a bad mistake. We arrived over the Larsemann Hills, which I erroneously though to be the Vestfolds, and this gave rise to later confusion.

On our return flight we flew over the Amery Ice Shelf just as clouds moved in beneath us. I suggested to Clemence that we should turn to starboard and move to intersect the coast but he said he was on the right course for the ship. However, I continued to worry and finally asked him to veer to starboard, arguing that if we could see the coastline we would be quite sure where we were. He finally agreed and altered course 30° to starboard. Shortly after that the cloud cleared away from beneath us and we crossed the coast to fly over the sea. It turned out that we had been well off course and, if we had continued, would have run out of fuel inland over the continent, somewhere well to the south of the Scullin Monolith! No-one would have had the faintest idea where we were.

The results of our flight I summarised thus:

- The Vestfold Hills (really the Larsemann Hills) were blocked in by fast ice. There was a group of islands further north of them.
- A new area of ice-free rock existed about 60 miles further north. This excited me for I thought we had discovered new land. (But it turned out to be only the Vestfold Hills!)
- Prydz Bay was hardly recognisable, a great piece of the Amery Ice Shelf, 50 to 60 kilometres long having broken off and drifted across to within 32 kilometres of the Vestfolds (Larsemanns).
- There were indications of a possible approach to Mawson to the east of the ship.

After lunch, pilot Doug Johnston and I set out to reconnoitre the approach to Mawson. I did not like the look of the weather. It was beginning to blow strongly from the east, with low cloud coming over us, but the OIC at Mawson, Jim McCarthy, said the weather there was all right, so we decided to go. We took First Mate Leo Christiansen with us.

Taking off at 1556 hours we flattened out below cloud at 215 metres but almost immediately we were forced down to 120 metres, and soon after, to 90 metres. It was a flight I remember vividly to this day—eerie, hazardous and strangely thrilling. Below was a ghostly, weird world of grey, black and white, the black sea studded with white ice floes like confetti, and visibility through the grey murk ahead only a couple of hundred metres except when snow showers engulfed us, in which case it dropped to about 50 metres. Shreds of wispy clouds whipped by to obscure the view further, while silent towering icebergs would loom up suddenly and sweep past close under our wings.

We were carrying mail for the Mawson men and were keen to deliver it; but, even more than that, we wanted to continue because we were enjoying the magic of it all.

At 1704 hours we were recalled by radio from the ship, which advised us that the weather was closing in. Our state of euphoria was such that we just laughed—we had been flying through such conditions for just over 45 minutes! Nevertheless, back we went and landed at 1736 hours. We had just taxied to the side of the ship when a hard, old, ice floe drifted in. The crew passed down a hook from the derrick, Johnston hurriedly attached it to the aircraft and we were lifted clear of the water just as the floe passed under us. The margin was about three metres.

From our reconnaissance it had appeared that there might be a chance of the ship pushing through to Mawson in a SSW direction, so I advised the captain accordingly and went to bed, quite exhausted. The ship set out through light pack and proceeded until 0400 next morning, when heavy pack forced it to stop.

I was called when the ship stopped, and decided on another reconnaissance. Johnston and I took off at 0600. We flew first SSE to look at a pool I had seen on our Prydz Bay flight, but found it was surrounded by fast ice. We returned to the ship and headed north then west, then south in a complicated path following the open edge of the heavy pack. We then set out for Mawson along what we considered to be a best sailing path for the ship via a large open pool some 27 kilometres long. I was busy sketching the ice and plotting our route while Johnston looked after the radio. Before we reached the pool he was out of range for radio telephone and had to resort to keying morse.

We reached Mawson, around which the fast ice was melting, and flew in low to buzz the station. Some of the men emerged with a red flag and waited on the rock above the station for us to drop their mail.

Johnston did a number of tight, banked turns, swooping down over the huts then zooming up steeply, while I tried (not very successfully) to photograph the station. He levelled out and approached on a long, straight run at about 30 metres to enable me to drop the mail. I struggled to open the door against the force of the slip stream and, with considerable effort, just managed to open it wide enough to push one hand out with the canvas mail bag. So that the bag would not hit the tail plane, I had to wait till Johnston banked, then saw the bag drop and the men run to retrieve it. I suddenly felt very ill and had to crawl back to my seat and lie back in order to avoid vomiting. I felt awful—clammy and sweating and nauseous—but luckily that was all. I told Johnston to return to the ship, entered details in the flight log, did some more sketching of the ice conditions then, when we reached the pack ice and there was nothing else to record, I lay back and went sound asleep until we landed.

I went on board, had breakfast then took my ice data to the bridge. I checked Ivan Grove's plot of our air track, sketched in the ice conditions and showed the result to Captain Hindberg. We decided to leave Mawson until later and go first to Prydz Bay. We had a long discussion about the ship's supply of bunker oil and what sailing we would need to do for the rest of the voyage. We finally decided to proceed to Prydz Bay to set up Davis Station, then to Vincennes Bay (Wilkes) to pick up some of our men from the US icebreaker *Glacier*. We would then return to Mawson to relieve the station, proceed again to Davis, then finally leave Antarctica for Kerguelen.

By midnight we were well on the way to Prydz Bay and the Vestfold Hills.

Next day Bruce Stinear said that he thought I had mistaken the Larsemann Hills for the Vestfold Hills. He did not convince me at the time, but the following day I found he was right when, as the ship sailed in towards the West Ice Shelf (north of the Vestfold Hills), Ivan Grove took me up for another reconnaissance. We flew over the ice shelf, then south over the Vestfold Hills and Rauer Islands to Ranvik Bay before returning to the ship. From Ranvik Bay we could see south to the Larsemann Hills and beyond. As we could see no islands north of the real Vestfold Hills, it was then obvious that what I had thought to be the Vestfolds on my long flight had been the Larsemanns. (The islands concerned were the Rauer Islands.) This meant that our flight path from the ship—intended to reach the Vestfolds but finishing up over the Larsemanns—must have been 30° off course. It was indeed remarkable that on both legs of our flight we had been about 30° off course, to starboard on the outward flight and to port on the return flight. I did not know whether the fault lay with the magnetic compass, or the astro observations taken to check it, or the pilot himself, and there was no way of ever finding out.

On 9 January, the ship proceeded through an accumulation of icebergs into open water about eight kilometres off the West Ice Shelf and it was from there that Grove and I made the flight above.

Next day we sailed west of an array of icebergs off the northern part of the Vestfold Hills and turned in towards the coast. When we arrived amongst the icebergs just NNW of Magnetic Island (where we had first landed in 1954), Grove took me up for another flight.

We flew over the Vestfold Hills, back and forth, for over an hour while I tried to measure up all the factors involved in choosing a site for our new station. Two seemed essential—a good approach for the ship and a water supply for the station. Other requirements were a good beach run-out for the DUKWs, reasonably level, clear ground upon which to erect huts, access to the plateau inland and clear horizons for meteorological and auroral observations.

We returned to the ship to pick up Bob Dingle and made a terrible landing. We bounced off the sea to a height of about 10 metres, then hit and bounced five metres, then hit a couple more times before settling down. It was a wonder the undercarriage survived it.

With Dingle on board we repeated the reconnaissance, but we both agreed that no obvious site was apparent, so we flew south to look at the Rauer Islands. there we saw a promising site at the extreme southern end of the group, on the mainland next to the glacier. The approach looked good and the glacier would provide water. Back at the ship we made another lousy landing, but not as frightening as the previous one.

At 1900 hours the ship sailed SSE to approach the Rauer Islands. When we arrived we found a strong katabatic wind blowing offshore. The water was very deep and at one stage the echo sounder gave no trace. The captain did not know whether the water was too deep or the sounder had developed a fault, so he dropped a hand line from the bows to 250 metres and also hung out an anchor on 90 metres of chain. Soon after, the trace on the sounder reappeared. If it was a chasm we had crossed it must have been a very deep one. We crept in very slowly, at 0100 hours on 11 January, towards the rocks of the mainland next to the glacier front at the extreme southern end of the Rauer Islands. Close in the water was too deep for the ship to anchor and the katabatic wind was blowing hard, so the captain turned about and slowly retreated.

I had asked for the ship's motor boat to be lowered so I could take a party ashore, but by the time this could be done, the ship had moved out a couple of kilometres. We therefore had to butt our way in through a choppy sea, against the katabatic, being lashed by douches of freezing water.

With me were the second mate (Frank Heidemann), Macey, Mather, Dingle and Stinear. We reached the shore at 0225 hours and made an abortive approach in a cove littered with rocks. Our second attempt was successful, where a gently sloping shingle beach provided what would be a good run-out for our DUKWs. I jumped out into a metre of water and waded ashore, followed by Dingle, Mather and Stinear. Heidemann and Macey stayed in the boat and cruised around just offshore.

We found ourselves on a moon-like landscape, rendered more desolate and forbidding by the dim half-light and the violent wind. A gentle rock slope, littered with loose rocks and boulders, led back to high rock cliffs, heavily banded with vertical strike. Access to the glacier snout to the south would be difficult and access to the plateau behind even worse. Although the ship could come in close, the deep water precluded a safe anchorage. There was enough clear rock space for the erection of huts but, obviously, water would be a problem. And from the aspect of the terrain it was clear that the site would be plagued almost continually by the unpleasantly strong katabatic wind. Further, the high cliffs ruined the site for meteorological and auroral horizons. I decided against it.

We climbed into the boat at 0330 hours and arrived back at the ship, pretty cold, just after 0400. I asked the captain to go back to the Vestfold Hills and we moved off at 0430. Then I fell into my bunk to snatch a couple of hours' sleep.

I rose again at 0630 and took soundings as we made an approach, hopefully, to the Ellis Fjord region; but it was very difficult from sea level to identify specific coastal features because of intervening icebergs and islands. Our first approach was wrong, for we reached a point just south of Magnetic Island, which I recognised. We turned about at 0900, headed south and came in again towards land. We anchored at 1045 hours to offload the aeroplane.

I had arranged for Clemence and Johnston to make a photo flight from the Rauer Islands northwards to the West Ice Shelf and they took off at about 1135 hours. They radioed back to us the heading we should sail to approach Ellis Fjord. The katabatic we had experienced was quite missing at our anchorage and the day was fine and sunny.

At 1245 the plane returned. A gauge had shown the oil to be overheating, but fortunately, it turned out that the gauge was faulty. De Havilland had been asked in Australia to clamp the thermocouple to prevent vibration but they had only taped it down, the tape had worn through, the piece had vibrated and finally it had broken. The plane had to be hoisted back on board while the airmen repaired it.

At 1300 hours the ship moved again, but ran into shallow water, and when it became less than 25 metres the captain put the motor boat out in front with a hand line for sounding and a walkie-talkie radio connection back to the bridge. We crawled along at one knot in the wake of the boat. I guessed we were about three kilometres offshore.

Our approach was no good, the depths ranging from 18 down to 9 metres, so we retraced our path, returned to roughly the place of our earlier anchorage south of Magnetic Island and anchored at 1650 hours. I reluctantly abandoned a proposed site just off Ellis Fjord that I had provisionally chosen from a US aerial photograph.

Clemence and Johnston finally got away at 1705 for a photo flight from Rauer Islands to the West Ice Shelf. After unloading the aircraft the ship attempted to approach Ellis Fjord from the north, setting off at 1845 hours and creeping forward in about 24 metres of water after a scare when a ridge rose up to 15 metres. At 1855 the ship anchored again and the motor boat went ahead to sound out a passage. I lay down in my clothes to rest—I had not actually been in bed since the night before the last one—but at 2000 hours the first mate came in and told me that soundings in front of the ship had varied 11-17-7 metres and that the captain had said that he was not prepared to go ahead. I didn't blame him! He had then gone to bed, worn out after two to three days on the bridge.

(Observations made in subsequent years showed that the entrance to Ellis Fjord was blocked by a submerged ledge of rock only one or two metres below the surface.)

The plane returned at 2100 and I met it in the boat and took off immediately with Johnston for another reconnaissance. I flew for 25 minutes and concentrated upon the stretch of coast opposite the ship and about five kilometres north of Ellis Fjord. I picked three inlets that I labelled 1, 2 and 3 as possible sites and three slight indentations on the coast (as possible fall-back choices) that I labelled A, B and C on a rough sketch that I made. It looked like this:

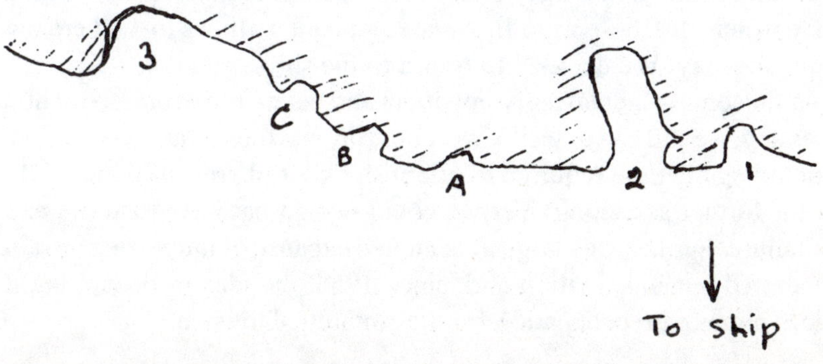

Landing at 2139, I collected Dingle, Macey, Stinear, Gallasch and Heidemann and set off in the ship's boat, with Heidemann driving, to examine the coast more thoroughly. The date was Friday 11 January.

We first examined area 1. There was no approach for the DUKWs—shallow water, many submerged rocks and rocky run-out at the beach. The possible station site was so rough with boulders that DUKWs would be able to deliver cargo only to the edge. Channels that appeared to have been cut by meltwater ran through the site, which also contained filthy sea-elephant wallows. There was also a seven degree cut-off on the northern horizon.

We pushed north to look at area 2, arriving at 2230 hours. There was a long approach of about half a kilometre of tidal shallows, mainly sandy but with a number of rocks near the beach. The distance from an anchored ship would be about two kilometres—too far. There was no solid rock base for huts, only glacial rubble and boulders, with few paths for DUKWs to pass between. A promising site on the north side of the inlet was spoiled by the approach over a rough rocky beach upon which ice floes were stranded. The nearby inland pools and lakes were salty.

We examined area A, closely but most of the above objections applied. It was a most unattractive place, so we headed for area 3. On the way we looked from the boat at B, but were not impressed enough to consider it worth while pulling in, mainly because it did not seem flat enough. There was a narrow beach with sea elephants wallowing on it and a low bluff blocking the approach of DUKWs to the flat ground higher up.

As we pushed on towards area 3 we realised that it would be too long a haul for DUKWs to operate economically from the ship, so at 1310 hours we turned back. On our return trip we looked in at area C. It had a good haul-out but no site approachable by DUKWs. There was a good horizon all around and a dirty snow drift nearby that, at a pinch, could provide water. But there were filthy sea elephant wallows just where the best hut sites lay. We decided to return to the ship.

I was becoming increasingly impatient and not a little depressed. Time was passing, we still had to visit Vincennes Bay and then return to Mawson, and including the time required to establish Davis, there was none left to spare for further searching. Yet we could not go back to Australia and admit failure. Further, this large unoccupied expanse of interesting terrain had captured our imagination and, since 1954, the idea of occupying it had been a major preoccupation for the Antarctic Division.

The eerie half-light at that hour made distant views indistinct and, in this strangely unreal environment, added to my depression. As the boat chugged back along the coast I decided, as a last desperate attempt, to have a closer look at B, so we ran the boat in and landed. It turned out to be better than we had thought earlier, but it still seemed far from ideal. Certainly there was a smooth sandy beach, uncluttered with boulders, which provided a good haul-out for the DUKWs. The distance from the ship was within practical limits. There was a low bluff behind the narrow beach, on top of which there was a flat expanse of boulder-strewn country. We climbed the 12 metre bluff and examined this. Most of it was so strewn with boulders as to be impassable for DUKWs but at the northern limit of the beach there was, above the bluff, a level stretch of ground covered with pebbles and pressed flat as though a steamroller had passed over it. It was obviously a snowdrift flat, compressed over many decades by the annual deposition of winter snow. The outlook to the horizon was good all around. There was one major deficiency—there was no water!

In winter, water would obviously be no problem, the evidence being that deep snowdrifts would abound. But the aspect of the surroundings was that of a gibber desert, so we would have to solve the problem of a summer water supply.

Another small problem faced us but it was not insurmountable. There was no access from the beach to the flat above the bluff; however, the escarpment at the northern end of the beach, below the flat, was composed of boulders, small rocks and rubble and not solid bedrock, so it seemed that some hard labour might gouge out a ramp to traverse it.

Because of our snow-drift theory of the origin of the flat it was pretty certain that any huts we built would be snowed over in winter; but they could be built in line and connected by passageways, and we had already designed the prefabricated huts with access hatches in the roofs

A consultation between Dingle (the OIC designate of the proposed station), Stinear (one of the members of his party), Gallasch (DUKW leader) and myself produced the unanimous decision that it was this site or nothing. It would of course be possible to put a station on one of the islands but such a move would have too many disadvantages and, anyway, time had run out for searching around.

We left at 1345 to return to the ship. On the way we took the boat directly offshore to where we envisaged our relief ships might anchor. When almost a kilometre from the shore, with the echo sounder running, we obtained depths of 15m, 12m, 16m and then, suddenly, 2m! I had just begun to worry about the captain's willingness to bring the ship close enough for economical DUKW operations (about half a kilometre was the limit) when I saw that the shallow area was obviously a shoal extending from two nearby islets. Beyond this line the water was uniformly 18-metres deep. A fairly large island (later named Anchorage Island) lay about 1.5 kilometres offshore and acted as a reference point for our later discussions with the captain.

Arriving back on *Kista Dan* at 0215 hours I had a hot drink then worked with Macey arranging work lists for the men. At 0330 I drafted a radiogram for my Minister, Richard Casey, in reply to the two which I had just received. The captain had arranged to rise at 0600, so I had a shower and rested for some time on my bunk. My thoughts were not wholly pleasant: I was immensely relieved at having found some sort of a site and at not having to return to Australia to admit that, in an area of more than 520 square kilometres, we had been unable to find a place to put our new station; but I was somewhat appalled by the difficulties I could see facing us and I was annoyed by the radiograms from Canberra.

I had previously sent Richard Casey a radiogram, following my first long flight to Prydz Bay, saying that Vestfold Hills was blocked by fast ice but that I had discovered a similar rock region further north and, if it proved suitable, we would establish the station there. (As already described, the fast ice region was the Larsemann Hills, the region north was the Vestfold Hills.) Casey apparently became worried and consulted Sir Douglas Mawson. He then sent me a radiogram relaying Mawson's advice, which was that the new area would not be as good as the Vestfold Hills but would do if the latter were inaccessible.

This was amusing on two counts. First, neither man had any idea of what the new land was like. Secondly, I was the only living person to have had first-hand experience of Prydz Bay, yet Casey sent me advice from Mawson who had never been there! I did not blame Mawson—he was just trying to be helpful. But I thought Casey should have had more sense.

At 0545 on Saturday the captain came to my cabin, so I described what we had found and asked him to move over near the island and as close in to the shore as possible. He did this and anchored about 800 metres out.

After breakfast things moved well. The aircraft took off on a photo flight, but cloud aborted the trip and the plane was moored on its floats aft of the ship. The DUKWs were hoisted out and sent ashore with 'barge' living caravans and camping equipment, while our 'Lollypop' launch was lifted off the hatch of the hold and put in the water alongside.

I went ashore in the first DUKW and landed at 1145 hours. The second DUKW followed with cargo and the launch with men. Gallasch chose a traverse slope for the ramp and a number of us began the heavy job of cutting out the track up the bluff. With crowbars we levered out the boulders and with picks, shovels and hands we cleared the small stuff. A DUKW winch helped drag big boulders away. After several hours' work we had a rough track to the top that the DUKWs could negotiate.

Meanwhile the two barge caravans and a tent had been set up as a temporary camp on the beach. A little further along the beach 40 elephant seals wallowed disgustingly in mud. Unloading began with two loads of fuel drums to clear the ship's deck, then hut sections came ashore. Dingle and I laid out the hut sites, in line on the flat along the top of the bluff. The aircraft carried out two photo flights including a mosaic of the whole Vestfold Hills area. We stopped work at 2200 and I went to bed for the first time in almost three days.

On Sunday 13 January the fine weather continued. All rose at 0430 and the first DUKW went ashore at 0520. It struck a rock near the water's edge when running into the beach and was stranded. The second DUKW had remained ashore overnight but its engine would not start so, eventually, the stranded DUKW was hauled off the rock by the ship's motor boat. I then set up marker flags on the beach to guide DUKWs in a safe approach.

We completed a great day's work. Number one between decks cargo hold had been emptied and a lot of cargo had gone ashore—hut sections, the two diesel engines and generators, the oil tank, the dinghy and timber. After dinner that evening the men finished erecting the sleeping hut—a record one-day accomplishment!

At 1600 hours I held a brief Opening Ceremony for the new station. All the men gathered around the half-erected hut, I raised the Australian flag and made the following speech:

This is an historic occasion. Our base here is the fourth Australian base ever established on the Antarctic Continent. We are indeed privileged to be members of the establishing party.

This station, the second that we have set up on the Antarctic Continent, marks a further phase in ANARE progress. It will be an interesting link in the network of IGY bases across Antarctica and its weather data will be of great value to Australia.

I now have the honour of naming this station. In the name of the Queen and of the Government of the Commonwealth of Australia I name this station 'Davis'.

The name is to honour Captain John King Davis, a famous Antarctic navigator and captain, First Mate Nimrod (Shackleton) 1908-09, Master *Aurora* (Mawson) 1911-14, Master *Aurora* (Shackleton Ross Sea Relief Expedition) 1916-17, Master Discovery (Mawson BANZARE) 1929-30.

At present he is retired, living in Melbourne, and is a member of the ANARE Planning Committee.

The names of Mawson and Davis are connected in Antarctic exploration. It is fitting that they should be remembered together as the names of our two present Australian Antarctic stations.

Rising at 0430 on 14 January we busied ourselves stacking food cases till breakfast time. After breakfast I flew with Grove to photograph the Amery Ice Shelf in response to a request from Melbourne by Leckie that we re-fly one of his earlier runs that had been unsuccessful.

It was an interesting flight of about five hours but disappointing in one respect. Because of lack of fuel we had to turn back just before reaching the place where, in 1955 on *Kista*, I had suggested that the shelf joined the continent. I would have liked to have checked on this from the air. Our flight path (sketched on the day) looked like this:

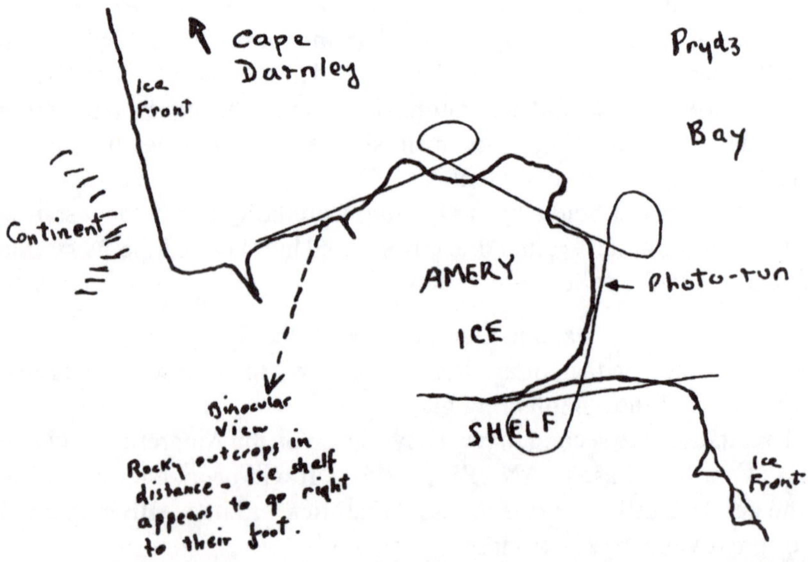

The men worked until 2230 and completed the community hut. All cargo was taken ashore, including 36 drums of diesel oil. We had to use the A-frame on one DUKW to unload these from the other, which slowed things down. There was still plenty to be done—build more huts, erect aerials and power lines, furnish huts, install engines and electrical generator and electric stove, tie down huts, etc. The chief engineer and his staff were constructing a still for distilling sea water. This would be the main source of summer water for the station.

Our fourth day at Davis was calm, sunny and enjoyable, but a proposed flight with Johnston and me to Sandefjord Bay had to be cancelled because the aircraft's oil temperature gauge was giving a low reading. It was infuriating that a perfect day for aerial photography was being wasted. I pulled the instrument apart and checked the thermocouple in ice and in steam and found it reading 12°—13° low. We then tested it by running the aircraft engine on deck until it was hot and found that the gauge read progressively lower and lower. I suggested that the pilots should fly and simply add 12° to the reading of the gauge but they declined and began searching in their spares for a new gauge.

The men ashore had a frustrating time. In the morning they could not proceed to lay the concrete floor of the engine room until the heavy diesel engine and generator had been lifted into place in a gap in the bearers; and those working on the auroral hut were held up because the bearer timbers had not come ashore.

The aircraft was hoisted out into the water and No. 2 hatch was raised. In the hold were the bearers for the auroral hut, which should have been in No. 1 hold. Also in there was the Ferguson tractor. Surveyor Fisher began an astrofix of the position of the station.

Work went on until 2200. I helped a gang complete the food stack and then, back on the ship, I worked on the aircraft temperature gauge, but without success. Bed at midnight.

Next day, 16 January, we started work as usual at 0530 hours. I worked ashore until breakfast at 0800, then decided to remain on board and catch up with clerical work; but I couldn't keep awake at my desk, so went ashore again and worked there until 2100. I calculated that I had had only 33 hours sleep in the last eight days.

My diary for the day noted:

This site improved on acquaintance. It was damn lucky that I decided as a last resort to make one more landing and look at a place that I had already passed once and classified as no better than the rest. In particular, the ship approach is almost a straight line from the open sea and we shall be able to go straight out without worrying about following our complicated inward passage. The landing approach for DUKWs has a few rocks, but now that we have our markers up we run in without trouble. The roads to the station and the fuel dump are now reasonable, thanks to all the work done by Lucas and Izabelle. The cleared sites for fuel are also OK. I was delighted to find a snowdrift flat, similar to the one our huts are on, just a hundred yards around the corner. If we ever have to extend the station, there is plenty of room if we clear the boulders from the intervening stretch of rubble.

I was not to know then that, by 1990, Davis would be our largest station in terms of men accommodated and scientific programs in train.

I informed the men that next morning there would be no work before breakfast and they could sleep in till 0700 hours.

To help solve the water supply problem, the DUKW men attempted, but without success, to tow a bergy bit in to the beach.

Work next day included the tying down of huts, completion of the engine hut, sealing huts with a sealing compound, fixing base plates and guys for the two Kelly and Lewis aerial masts, and completing the dome and installing equipment in the auroral hut.

On the day after that Macey finished erecting the two Kelly and Lewis masts and two smaller wooden ones, with their accompanying aerials. The ship's radio was still trying to contact the US icebreaker *Glacier* with which we had to rendezvous near Wilkes. I received a nice radiogram from Captain J. K. Davis in reply to one I had sent him when we named the station. A congratulatory radiogram was also received from Hans Christian Petersen, captain of *Magga Dan*.

Physicists Mather and Goodspeed walked a 20-kilometre circuit carrying out a gravity survey. I took the obligatory photograph of the wintering party and then searched for a source of water reasonably near the station. I found a pool of brackish but drinkable water about a kilometre away, overlooking the northern edge of the inlet I had called 1. This relieved my apprehensions about a water shortage in case the distilling apparatus should fail for any reason. The DUKW crew hauled in a bergy bit and stranded it in shallow water where the dinghy could reach it. Ice chopped from this would provide an auxiliary water supply.

Our ninth day at Davis dawned fine and calm. We had had a fantastic run of good weather for unloading and building. I arranged with the captain that we should depart at 1000 hours, so work began at 0500 loading everything that was to be taken away. I rose at 0400 in order to make use of the fine morning to make the flight to Sandefjord Bay that I was so keen to do. The region had intrigued me even since seeing Roscoe's map compiled from the US 'Operation Highjump' aerial photography.

Johnston and I took off at 0544 and the warm glow of the low, slanting, early morning sun delineated the relief of the features of the panorama before us to perfection. I took numerous photos but knew that they could not hope to reproduce the glory of the scenery.

At the bottom of Prydz Bay the plateau ice from the south and SSW disgorges in the form of the giant Lambert Glacier to form the Amery Ice Shelf. Ice from the south and SE flows out into the sea from the Polar Record Glacier and others to create the Publications Ice Shelf. Between these two shelves lies a confusion of jumbled ice from which protrude the Munro Kerr Mountains. It is a very difficult region to map, not only because of the problem of interpreting the confused jumble of ice forms but also because the configuration of these forms changes from year to year as glaciers move and icebergs break off from the edges of the ice shelves.

My diary records my observations thus:

Polar Record Glacier could, I think, be called a glacier but the ones next to it on either side are not; they are merely ice-falls where plateau ice tumbles over cliffs of the underlying rock. Roscoe's Polar Record Glacier Tongue is, I think, straight ice shelf, not glacier tongue. Sandefjord Bay is a jumble of tilted bergs and ice shelf all blocked up with built-up snow. A line of rock outcrops runs SW from the ice-falls mentioned above and I took a panoramic spread of photos of them on my Leica camera. The sea level ice shelf runs right to the edge of these and continues on into the distance. I could not see the Prince Charles Mountains. The whole area needs redrawing very differently from that shown by Roscoe.

We flew over Lichen Island which a party of us had visited in 1955. It—would not be possible at this stage to do the trip we did because of the broken up fast ice and numerous cracks and wide water gaps. The edge of the fast ice seems about where it was in 1955.

We flew over the Bolingen Islands to the coast then turned north and flew home taking shots of the Larsemann Hills but, like Rauer, they emerged steeply from the sea and landing places are hard to find. One large fjord (about second from the northern edge) seemed open and navigable and landing places appeared to exist on its southern shore. The fast ice that I saw some 10 days ago on my flight over Larsemann from Mawson was gone.

We landed beside *Kista* at 0616 and were hoisted aboard.

The ship duly sailed at 1000 hours, heading for Wilkes. The station at Davis had been successfully established. A pioneer party of Dingle, Stinear, Lucas, Lied and Hawker had been left to winter there and consolidate the bridgehead as a base for further development. The whole of the eastern edge of Prydz Bay had been photographed, as also had the terminal edge of the Amery Ice Shelf. All in all, it had been a highly successful exercise.

19

AN ERROR OF JUDGMENT

In 1961 MV Magga Dan *narrowly escaped a disastrous collision with an iceberg. I have experienced only one dangerous incident involving a ship and an iceberg and it occurred because of stupidity.*

On 23 February 1961 we had just completed the construction of an automatic weather station at Chick Island and had set sail for the French station at Dumont d'Urville. We crossed a large bay whose surface was freezing into black 'new' ice and moved north into an accumulation of icebergs. These were part of the western edge of the Dalton Iceberg Tongue which jutted out about 65 kilometres from the coast just east of Chick Island.

When the sea starts to freeze, one first notices myriads of tiny ice crystals forming at the surface. These thicken up until the surface has an 'oily' appearance. This is called 'frazil ice'. As the crystals coagulate, they form a thick, soupy layer on the surface called 'sludge'. Further aggregation produces plates like large waterlily leaves on the surface and, as wind and water action moves these around, they develop turned-up edges through bumping into other plates. Such ice is called 'pancake ice' and pancakes may be anything from half a metre to three metres in diameter. As the pancakes freeze together and produce large thin areas it is called 'new ice'. If new ice forms in still water from sludge without producing pancakes, or if the pancakes produce new ice without snow falling on top, the new ice has a black appearance, because it is thin enough and transparent enough for the dark water beneath to show through.

At midday our ship, the *Magga Dan*, became stuck while Captain Bill Pedersen was attempting to pass close by the western edge of an iceberg. It was just bad luck: one minute the ice was reasonably slack and the next minute the pressure was on, with ice cracking and rafting up all around us and building up into ridges elsewhere.

Bill and I made a helicopter reconnaissance flight at 1250 hours with pilot John Stanwix, but we stayed up only six minutes. We saw black ice all around except for the 100-metre stretch in which we were stuck. If we had gone the other side of the iceberg there would have been nothing to stop us for the next 25 kilometres!

The crew unloaded one anchor and its chain onto the pack ice and the engineer transferred 50 tonnes of fuel from the forepeak tanks to the bottom tanks to lighten the ship up forward. Two ice anchors were then set out aft of the ship, and hauling on these with the winches and with engines full speed astern, we slid backwards and were free. The captain then spent some time turning the ship. At 1525 hours we proceeded north, along the western icebergs of the Dalton Tongue, through new ice 15 centimetres thick now under considerable pressure from a stiff SW wind.

For some time all went well. The captain charged a large bay ice floe standing a metre above sea level and barged right through it, with the ship rearing and bucking and cabin furniture being hurled around.

On the way in to Chick Island, I had made a helicopter reconnaissance and had noted a 'bay' of open water relatively free of pack ice on the western edge of the iceberg tongue, ringed by five large icebergs. As the SW wind was beginning to build up thick pack along this edge of the northward-trending tongue, I thought it a good idea to shelter in this bay until the wind went around to the east and loosened the pack. Inside this protected area there was new ice but no pack. I suggested this to the captain, who concurred.

The move involved swinging the ship to head east into the 'bay', taking advantage of some fairly loose pack close to the large berg dead ahead, which was the northern 'doorpost' to the opening. Bent Hansen, the first mate, was navigating the ship from the crow's-nest on the foremast but, to my consternation, he directed the ship towards the centre of the gap leading to the bay, across which was a dense strip of pack ice about 400 metres wide. It was odds-on that we would stick in the middle of it, so I asked the captain to instruct Bent to head for the loose ice close to the large berg. The captain said that Bent could see the patch of loose ice near the berg and would no doubt steer for it, so declined to intervene. (It would have meant using the 'intercom' telephone from the bridge to the crow's-nest.)

But Bent went straight into the pack ice and, sure enough, we had penetrated little more than a ship's length before we were brought to a halt. It took us some time, using all our power, to back out inch by inch.

The captain then told Bent to sail around where I had indicated. He had a large area of pretty open water to port and I expected him to take the ship out into this, then line up due W-E and take a straight easterly course, parallel and close to the southern edge of the iceberg, through the loose ice.

But no, Bent headed north then began a broad sweeping turn aiming to swing to the east and pass close by the berg on a tangent. However, he misjudged the radius of his curving path.

Bill and I were standing together on the bridge. It was obvious to us that Bent's curved path would intersect the berg, but we assumed it would be obvious to him also, and we waited for him either to turn more sharply to starboard or else to put the engines into reverse. There was a distance of about 400 metres to go.

But on we went with speed undiminished until, suddenly, Bill and I yelled at each other while Bill sprang and reached for the engine speed controls. At the same time Bent, at last realising the danger, put the engines full speed astern while swinging the rudder hard to starboard. But a ship under way is hard to stop, even with engines full speed astern.

On we went, decreasing only slowly from our three-quarter speed, and the iceberg loomed closer and closer. While we held our breaths the iceberg, with crumbling faces rising 20 metres vertically from the sea, made a tangent to our curving path and we slid along past it at about six knots at a distance of not more than five metres, expecting the vibrations of the engine to bring hundreds of tons of ice crashing down on our decks. Some fifty metres on we came to a stop, then backed away to the accompaniment of the exhalation of breath from all the spectators.

20

ON BOARD MV *NORSEL*

On the evening of 28 December 1949, I sailed from Cape Town for Antarctica on board MV Norsel, *as Australian Observer with the Norwegian-British-Swedish Antarctic Expedition (NBSAE).*

I had arranged to accompany the *Norsel* in order to gain experience of Antarctic expeditionary activity in preparation for organising, hopefully, at some future date, an Australian expedition to the Antarctic mainland. The voyage proved to be of immense value to me; I could not have chosen a better venture upon which to serve my apprenticeship. Of the total experience, the most valuable for me in later years was what I learned about navigating a ship in pack ice. The *Norsel,* for its size, was a fine ice ship and her master, Captain G. Jacobsen, was highly experienced in ice navigation in Arctic waters.

The aim of the expedition was to establish a station somewhere on the east coast of the Weddell Sea. This sea provides some of the most difficult ice conditions of any part of the Southern Ocean and the *Norsel* spent more time pushing ice in one summer season than any other ship I know of. By the time we had arrived back at Cape Town, on 7 March 1950, we had sailed 3600 kilometres through the pack ice zone, 2400 of these through heavy pack ice. During the voyage, an incident occurred on *Norsel* that illustrated several aspects of what we call 'luck'.

The food on the ship was frugal—at the minimum level required by Norwegian law for catering for seamen on sealing vessels. We were hungry for most of our time at sea. When we entered the pack ice and began collecting seals for dog meat and biological specimens, we started a routine of cooking a supplementary meal for supper each night. After dinner, when the cook and stewards had cleaned up, we would select some viands from the seal carcases—brains, liver or steak—and then, between 2200 hours and midnight, we would cook these on one of our field equipment petrol stoves in the galley.

On Sunday 8 February I cooked fricasseed brains for several of the expeditioners, then retired to my cabin to write. I worked on until about 0130 hours and then went to bed. I was awakened about 0530 hours by a terrific boom, followed by a bedlam of crashing and smashing. Our cabin door flew open, the air filled with dust, and splintered wood and articles crashed down from shelves and bulkheads.

My first thought was that the ship had struck an iceberg, so I sprang out of bed and looked along the passage, into which pyjama-clad men were stumbling. But no emergency alarms sounded and there were several minutes of confusion while men gathered in groups and tried to find out what had happened. It soon transpired that there had been an explosion of some sort.

Whatever the cause, the damage was considerable. All the doors in the accommodation part of the ship had been blown open. The heavy mahogany door of the main saloon on the deck above us had been split from lock to architrave, doors of cabins near ours had had their locks ripped out, all the glass in the cupboards in the officers' mess had been shattered, half a wall panel had been blown in Walford's cabin, and the pantry door opposite the galley was lying flat on the deck. There was dust, splintered wood and scattered debris everywhere.

It transpired that Fred Roots and a seaman called Felix had been on rostered duty ice-watch on the bridge and, when their time had expired at 0400, they had gone to the galley to cook their seal-meat supper. Their petrol stove was standing on the cold galley oil-fired range. When their meal was cooked, they turned off their stove and sat at a table to eat.

A few minutes later a seaman came to the galley to light the big stove so that it would be ready for the cook at 0600. As Roots and Felix were finishing their meal, they suddenly thought, 'What about our petrol stove?' and, simultaneously, they both sprang to grab it from the top of the hot range. As they reached for it, it exploded.

Neither man was severely injured. Standing almost over the stove, Roots had a singed beard and a slightly burnt face, while Felix suffered a burnt hand and a singed waist to his jacket.

The men were victims of both bad luck and good luck. The bad luck consisted of the unusual combination of circumstances and timing that led to the explosion, while the good luck was their escape from serious injury.

Apparently, the main damage from such an event is caused by the momentum of a mass of air rushing outwards from the seat of the explosion. The men were standing so close to the stove that the mass of moving air was quite small, although the heat was intense. Further out, where a much larger volume of air had been affected, the force exerted was much greater.

21

FIRST LANDINGS AT LARSEMANN HILLS

The establishment of Law Base at Larsemann Hills early in 1987, followed by bases set up by the USSR and China, lends interest to the following account of the pioneer ANARE landings there in 1958. The events occurred on the voyage of exploration that I led in MV Thala Dan *(Captain K. Hindberg) during the International Geophysical Year.*

Early in 1958 *Thala Dan* had called at Lewis Island, Wilkes Land, to establish an automatic weather station; had made an abortive attempt to reach Oates Land; had visited Dumont d'Urville (French Station); had picked up my administrative officer, Jerry Donovan, from Wilkes (USA) using our Beaver aircraft (because our ship could not penetrate the pack ice in Vincennes Bay); had visited Mirny (USSR); and had relieved Davis Station.

The date was 7 February 1958. As we were a couple of days ahead of schedule, I decided to attempt a landing at the Larsemann Hills (about 60 miles south of our Davis Station), which no ship had previously succeeded in approaching.

The first persons ever to set foot upon the Larsemann Hills were members of the 1957 ANARE wintering parties. Flight Lieutenant Doug Johnston, RAAF, flew the expedition's Beaver aircraft (A95-202) from Mawson to Davis, with Morrie Fisher, the Mawson surveyor, as passenger. From Davis, Johnston ferried Fisher and two men of the Davis party—Nils Lied (meteorologist) and Bruce Stinear (geologist)—to the Larsemann Hills, landing on skis on the frozen sea on 12 August 1957. Leaving the three men camped in tents on an island, Johnston flew back to Mawson.

Blizzards, either at Mawson or Davis, prevented the return of the Beaver for some time, and the three campers were finally picked up by

Flight Lieutenant Peter Clemence on 24 August. During their stay, Fisher obtained an astrofix on the island and the men walked over most of the mainland rock area.

However, back to 1958. We ran down Prydz Bay, first through open pack, then in open water in perfect weather, recording depth soundings as we went. By noon we were closing in on Larsemann Hills, visible as light-brown rocky features standing out against the white backdrop of the Antarctic plateau ice. I asked Ian Adams (1958 OIC for Mawson Station) to organise a shore landing party of about 12 men in two Army DUKWs with survival equipment and food for four days. I then decided to prepare, also, a motor launch team with Jerry Donovan (my lieutenant), Osmar White (journalist), Blair (mechanic) and myself.

The launch had been built with money donated by MacRobertsons Ltd, chocolate manufacturers. I had designed it for use in ice and had ordered it from the well-known boatbuilders, Halvorsen of Sydney. Of 7.5 metres length, with a powerful inboard diesel engine and an endurance of more than 450 kilometres, the boat was extremely seaworthy and had been sheathed with greenheart oak to protect the hull when working in ice. The boys nicknamed it 'Lollypop'.

I had planned, when we anchored off Larsemann Hills, to send the Beaver aircraft, mounted on floats, on a photographic flight to the Lambert Glacier and mountain features at the bottom of Prydz Bay. However, as the sea was calm and the weather perfect and as it looked like being late by the time we arrived at Larsemann Hills, I decided to start the flight immediately. So the ship stopped, the plane was hoisted out and Sandercock took off at 1450 hours. The ship then proceeded towards the coast.

The approach was reasonably free of pack ice and we were able to sail right in close to a large island about eight kilometres north of what appeared to be a major fjord cutting south through the Hills. The ship did not anchor; it simply dropped the two DUKWs and the motor launch and sailed back several miles out to sea again. This over-cautious act on the part of the captain surprised me, but I did not remonstrate. I took the wheel of the launch and led the vehicle convoy southward into the fjord, but all the western side was unapproachable because of fast ice or else too steep for a landing. A southerly wind sprang up and began to drive ice floes, breaking from the fast ice, across into the bays on the east side where we hoped to land.

I did not like the idea of leaving men ashore under these conditions in case the ice prevented our picking them up again later. However, after some searching, we found a reasonably sheltered cove of ice-free water on the eastern side and the DUKWs landed the men with their instruments and survival gear on an ice ledge above high tide level. There was no place where the DUKWs could run ashore, so I sent them back to the ship.

My launch party then set off south to see what we could find. First, we discovered that a row of islands on either side of us had given the impression that we were inside the fjord; now we could see that the mouth of the fjord was still ahead. We had landed our DUKW party on a large island on the east of this corridor.

A short distance further, on the west side of the corridor, there was a large, high island and we could see a rock ledge at water level beneath high cliffs. There was a strong tidal rip and the approach looked shallow, so, rather than risk the motor launch, I asked Jerry to take me in the launch's dinghy and put me onto the ledge.

From the ledge I made a hard but not too difficult rock climb up a 20-metre cliff, then scrambled a few hundred yards to the summit of the island. From here I took a round of photos and enjoyed a panoramic view of the mainland close ahead, with the broad fjord leading on towards the plateau ice. The sea ice situation in the fjord and amongst the islands did not look good, with ice floes and sheets broken from the fast ice drifting around to produce a continually changing situation.

I wasted no time returning to the boat and we sailed straight back to the ship, arriving about 2000 hours. At about the same time the aircraft returned and was hoisted on board after landing luckily in the only open water left around the ship. I swallowed a quick dinner, then returned with Blair in the launch to pick up the shore party. As the DUKWs were not designed to withstand impact with ice, I could not risk them in this situation.

When we reached the cove we found it filling rapidly with sheets of young new ice, 20-30 centimetres thick. The shore party were waiting anxiously on the ice ledge with all their gear, apprehensive about our prospects of evacuating them. I used the launch at full power to push some floes aside and one or two I managed to crack by charging them, but then they jammed against the shore and would not budge.

I had difficulty turning the launch to retreat before it became trapped

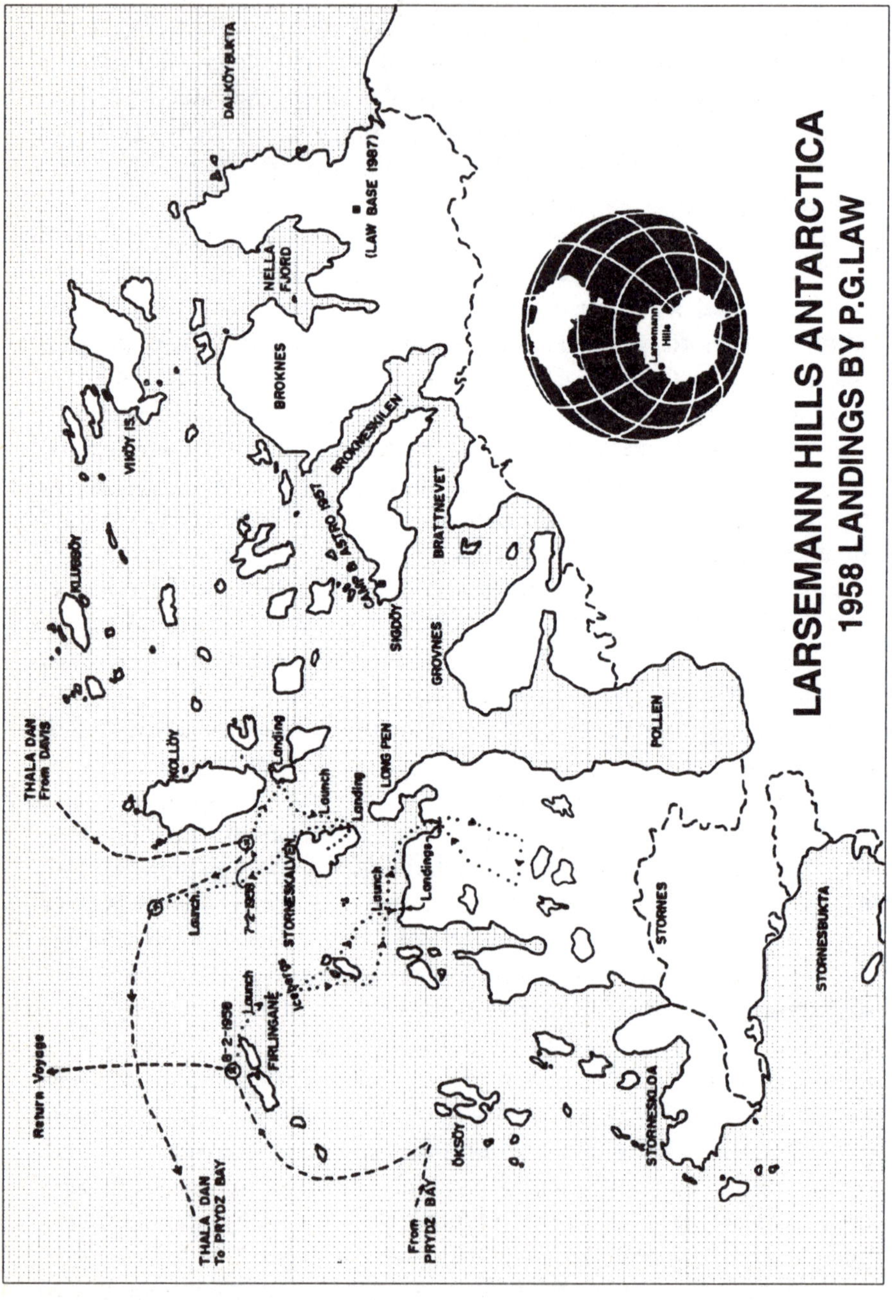
LARSEMANN HILLS ANTARCTICA
1958 LANDINGS BY P.G.LAW
DALKÖY BUKTA
(LAW BASE 1987)
NELLA FJORD
BROKNES
BROKNESKILEN
VIKÖY IS.
KLUBBÖY
BRATTNEVET
SIGDÖY
GROVNES
POLLEN
KOLLÖY
THALA DAN From DAVIS
Landing
Launch
LONG PEN
STORNESKALVEN
Landings
STORNES
STORNESBUKTA
FIRLINGANE
Return Voyage
THALA DAN To PRYDZ BAY
From PRYDZ BAY
ÖKSÖY
STORNESKLOA
Larsemann Hills

but, when I had done so, Jerry plied the dinghy back and forth to ferry the men and their luggage to the launch. We had them safely back at the ship and the launch hoisted by 2145 hours.

The landing had been too late in the day for Knuckey to get an astrofix, but the party had done useful geological work and magnetic observations.

The RAAF photo flight had been only partly successful. Someone, at some time during it, must have knocked the camera switch, for only about the first 30 minutes of a 150-minute photo run had been recorded. I thought it pretty careless of the observer not to have noticed that the cameras weren't running!

I decided to push deeper south into Prydz Bay, hoping that ice conditions around Larsemann Hills might have improved by the time we returned. So I asked Captain Hindberg to proceed towards the Bolingen Islands, which were the most southerly group in Prydz Bay.

We went west into ice-free water, then headed south. At almost 2300 hours we met an amazing array of icebergs. There were hundreds, of all sizes and shapes, many jumbled together by tides and storms, tilted fantastically and rafted one on another to from an impenetrable barrier. In the dim half-light at midnight, Jerry, Osmar and I stood on the bridge, tired after a big day but unable to break from the fascinating scene. The evening was calm and the sea smooth and the areas between the icebergs were covered with thin sheets of young ice and the oily shine of newly freezing frazil ice crystals.

I recalled man-hauling a sledge the year before over solid fast ice to find a path through this strange graveyard of bergs and being fearfully impressed by their awesome majesty and potential menace.

The ship reached a point about five kilometres from the Bolingen group but then was blocked by fast ice. We could have broken through laboriously by backing and charging, but we could not afford the time it would have taken. I also decided, in view of the ice conditions I had seen in the fjord, that it would be unwise to send a party ashore over the fast ice, which at this time of the year was likely to break out at any time. So I told the captain to return to the Larsemann Hills. Captain Hindberg again showed remarkable timidity. If the depth became less than 100 fathoms he would reduce speed to dead slow. Once, when the bottom shelved up gently from 100 to 60 fathoms, he signalled full speed astern as though we were about to run aground!

Just after midnight we were approaching a promising land feature that was free of fast ice and either a part of the mainland or a close island, but when the depth declined to 20 fathoms, he refused to proceed further. With 36 metres of water beneath his keel, I considered he could have edged forward at dead slow speed. I seethed with frustration but said nothing. After backing off he went back to the vicinity of an off-lying island, dropped anchor, and said that I had better use the motor launch if I wished to approach closer. So I called Jerry, Osmar and Jim and we set off in the motor launch at 0115 hours on 8 February.

It was eerie in the twilight and moonlight with a dull reddish tinge on the sunset—sunrise horizon in the south. We chugged our way past two islands, as I wanted to reach the mainland if possible, and pushed on through narrow chasms of blue space between towering icebergs in water whose surface glistened with the oil-slick appearance of frazil ice. The temperature was about 18°F. I kept a mental note of the appearance of various bergs so that we would be able to find our way back.

One iceberg in particular I photographed which rose to a towering pinnacle whose sides gleamed with light reflected from numerous angled facets. Beyond it we pushed through fast-freezing sea that already had an inch of sludge ice on top, dodging floes of young ice more than six inches thick and as hard as rock. A kilometre on there was another long island with two smaller ones to the left, one of which rose to a knoll about 60 metres high. On the rocky shore of the smaller one I was surprised to see spray flying and noticed that a ground swell, with a tide race between the islands, was swirling and dashing on its western side. We edged cautiously through the narrow gap between the large and small islands, feeling very confident in the splendid launch. This gap looked nasty as a possible ice trap on the way back. However, we took the risk and, once through, saw the rocky mainland about a mile further on—the major peninsula of the Larsemann Hills.

I was disappointed to find, as we approached through heavy sludge ice, that the coast was fringed with several hundred yards of fast ice. We would have to risk proceeding across this by foot if we wanted to get ashore.

So we tied up against the edge of the fast ice and left Blair in the launch with the motor running. Osmar, Jerry and I clambered out onto the slippery, polished surface of the ice and walked towards the shore. We noted a number of cracks in the fast ice up to 15 centimetres wide, which did not surprise me because of the swell I had noticed earlier.

Nevertheless, it was clear that we could not afford to stay ashore for very long.

Conditions at the tide crack fringing the rocky coast were nasty. The sea ice cracked and heaved, black water hissed and gurgled through the fissures and pulverised ice, under pressure, was forced into irregular ridges. We hopped and jumped through this belt to reach the rock, fearful of a leg being pinched between grinding floes. Time was even more pressing than I had thought.

I told the others to only stay long enough to say that they had been ashore and then to start back towards the boat. I then dashed ahead to a higher rocky area from which to take photographs. The light was poor but even bad photos of a new coast are better than none. This took only a few minutes and I raced back, crossed the tidecracks with a series of jumps, and hurried after my three colleagues.

We were not a moment too soon. The whole area of fast ice was rapidly breaking up and drifting out. What had been a 200-metre approach became a 600-metre retreat, by a tortuous course, over drifting slabs of ice, detouring around areas of open water and jumping the narrower gaps.

We found that the launch was no longer moored in open water to a continuous edge of static ice. This edge had broken up and the drifting slabs had threatened the boat. Blair had been forced to keep moving to avoid them. The launch now lay in a narrow channel between floes whose configuration was constantly changing. Blair waited anxiously as we chose floes as 'stepping stones' to approach the boat. By the time we got aboard, the channel had become blocked by ice and I had to use the full power of the launch to break one floe and to push others aside before we emerged into open water. We carried a deep groove in the starboard side of our wooden hull as a result.

We retraced our path and found that the gap between the islands was, as I had foreseen, becoming choked with ice. We managed to push through but it was clear that I would not be able to land my scientists on the mainland by this route and would have to settle for one of the larger islands.

Our path to the ship was through thickening sludge ice, for the temperature was 19°F and the sea was freezing fast. We went on board at 0310 and I called Bruce Cook and Eric Jesson (geophysicists), Graham Knuckey (surveyor) and Ian McLeod (geologist) to make an island landing in the launch with Blair and me.

While they were collecting their gear I wrote a hurried note for the airmen, detailing a flight I required to repeat yesterday's unsuccessful photographic run to the Lambert Glacier. I also instructed the captain to take the ship further out to more open water suitable for the plane's take-off.

I might mention here that pilot Bill Wilson had just taken off for a photographic flight and climbed to cruising height when his propeller speeded up abnormally. Apparently some fault had occurred in the hydraulic system that adjusted the pitch of the propeller. He was forced to land. Fortunately he was still close to the ship. He later got away successfully.

At 1530 hours my launch party set off. When we reached the iceberg I had previously photographed, I found that drifting ice floes had blocked all except a 3.5-metre gap hard against the tall pinnacle. I wondered, first, if we went through would we be able to return and, second, what was the chance of our engine vibrations causing the pinnacle to crash down onto the launch? These were matters of prime concern to the others, so I stopped the launch and asked them were they prepared to take the risk of proceeding. They agreed, so I went full bore through the gap while we all held our breaths.

The sea between there and the island I had chosen for our work had frozen further to the pancake stage and the engine was working hard to push the launch through it. There was a constant cracking, swishing and scraping noise as the wooden bows parted the solidified ice. We left behind a wake of broken panes like plate glass.

I was quite worried, because if the men were to stay ashore, as planned, for nine hours in order to get an astrofix and do magnetic and geological work, the ice could freeze up beyond the five-centimetre thickness that the launch could handle. Should we risk it? I reckoned that if any wind arose it would break up the new ice, so I decided it was worth a chance.

We moored to an ice ridge attached to the island above tide level and unloaded the observation party with survival gear and instruments. Blair and I then sailed back to look at the iceberg gap. The new ice was even thicker. However, on looking around, I saw some open water about half a kilometre to one side, so I took the boat over to investigate. I discovered a roundabout route around the far side of another set of bergs that led through more open water back to the ship.

I returned to the island, much easier in my mind, to find the men busy with their observations. The temperature was 24°F and a wind arose that made it bitterly cold on the crest of the hill where the two surveyors had set up their theodolite and from which I took a round of photographs.

The rock of the island was the usual dark brown granitic gneiss, with a lot of cream- coloured felspar and much garnet. Clusters of garnet and black granite were common as blisters on a lighter coloured gneiss and there was a lot of black dolerite like the rock of the dykes at the Vestfold Hills. The whole of the rock was heavily weathered, with cells eroded by blown sand and snow, and in one part on the eastern side the wind had eroded two parallel gulches rising up the flank of the island in a NE direction. In these were the usual patches of deposited salt from wind-blown sea spray.

Snow petrels were nesting on the island but, apart from circling skuas and fluttering Wilson's petrels, we saw no other birds.

The men built a rock cairn on the summit and beneath it they placed a glass jar containing a note that I had written giving details of our landing. (This note was found by geologists working from Law Base early in 1987.) From there I looked across a stretch of water to the east and saw the island whose steep coastal cliff I had climbed the day before. We had come back to within five kilometres of my previous landing!

I spent an hour photographing the men's activities in cine film. Then I picked up Jesson and McLeod who had finished their work, and with Blair set out for the mainland at 1030 hours. On the way I recorded depth soundings using the launch echo sounder.

During the morning the wind had blown all the newly formed ice and old floes across to pile up against the iceberg gap we had used, but my alternative route back to the ship was open so I did not worry.

When we reached the northern extremity of the main Larsemann peninsular we landed on a ledge of ice above which, a few hundred metres inland, arose a large bluff with interesting geological strata. As far as I could determine, our mainland landing place early that morning had been 1.5-2 kilometres further west.

The coast was now quite free of fast ice. Between 0200 hours and midday it had all broken up.

We walked to the bluff beneath which was a lake with interesting green algae on the bottom, giving the lake that emerald-green hue described by the Americans for the Bunger Hills lakes. We climbed the bluff and saw, beyond it, a hill about 200 metres high which we also climbed. From there the view was magnificent. I took photos and made a rough sketch of the panorama.

Below it, on all sides except the north, lay lakes. One very large one was completely frozen, in contrast to the others.

We returned to the launch at 1245 hours and retraced our steps to pick up the men remaining at the island. From there we made a wide detour through magnificent icebergs to follow open water back to the ship. The launch was hoisted on board at 1435 hours and at 1500 hours we set sail for Mawson.

Knuckey had fixed the cairn site on the island as being 69°23'S, 76°03'E. This fix, together with that obtained by Morrie Fisher the year before, would pin down the southern end of the photographic flights along this coast, while the northern end would be accurately determined by the position of Davis at the Vestfold Hills.

I spent until 2200 hours writing an account of our operations and then, after showering, went to bed for the first time in two days. We had ‘cracked’ the Larsemann Hills, a feat in those days (without helicopters) of no little difficulty.

22

BY PONTOON TO HURD POINT

In December 1953 we chartered Kista Dan *for the first time. We planned to relieve the 1953 wintering party at Macquarie Island and to replenish the supplies there. After that, we would return to Melbourne, reload and proceed to Heard Island, Iles de Kerguelen and Antarctica.*

In 1952 we had decided to supplement the food supplies at the island stations by sending down live sheep which the men could slaughter from time to time to provide fresh meat. At that time we had neither ship refrigeration nor station refrigeration, so all meat had to be either canned, salted or dried.

The first consignment of sheep went to Heard Island in 1952 on the little Norwegian sealer *Tottan*. We had designed strongly made sheep pens which, in Melbourne, had been secured on the foredeck of *Tottan* and we had placed in them 24 Border Leicester sheep. We had left Melbourne on Saturday 9 February and on the Sunday afternoon the weather had deteriorated. That night, at 2230 hours, the ship was butting her way into a Force 6 NW gale and her bows were pitching up and down through an arc of about six metres.

In the early hours of the morning I was awakened by the first mate who said, 'The sheeps—they is loose'. I scrambled out of my bunk, asked him to call Dick Thompson and Lem Macey, pulled on a parka, trousers and gum boots, and went out onto the dark, storm-lashed deck.

The freeboard of the *Tottan* when fully loaded was only about 70 centimetres, so when she rolled the waves came through the scuppers or tumbled 'green' over the bulwarks. On this occasion a large wave had struck the ship on the port quarter and had crashed over the deck, smashing the sheep pens to pieces.

The ship was pitching badly and rolling about 20° each side of the vertical. The decks were flooded as the waves came over the sides on each roll and the poor sheep were washing back and forth on the deck between the scuppers and the hold hatches in amongst the wreckage of the wooden pens and the straw on which they had bedded down. In the dim light from the lamps on the bridge wing it looked a shambles. My heart sank.

With the help of the crew, Thompson, Macey and I, staggering around in waves up to our knees on the rolling deck, first pitched all the broken timbers overboard. We then held a hurried consultation with the mate about where to put the sheep. They obviously could not remain on the foredeck. We finally decided to spread straw on the wings of the bridge and to take the sheep up there. The mate ordered the ship to heave to, after which she continued to pitch steeply but not to roll. Thompson roused out a few more helpers, and we began the heavy job of manhauling the sheep. First, they had to be hauled along the alleyway on the port side of the main deck accommodation; then they had to be lifted over a 50-centimetre steel section blocking water from entering the main passage in the accommodation area and up two narrow companionways, then finally through the bridge proper out onto the wings.

I remember it all as a crazy, unpleasant nightmare. The sheep were frightened and, with their long fleeces saturated with water, so heavy that they could hardly stand. We used two men to each sheep, one at the forelegs and one at the hind legs. I recall being at the rear and trying to push a sheep's rump up the companionway while Thompson spat and cursed as he hauled on the forelegs from higher up the ladder. My face was buried in the dripping wool of the sheep's tail and the stench of wool, lanoline and dung was sickening. We heaved and lurched as the ship rolled, and exerted all our strength to hold and raise the kicking animal.

As each sheep reached its destination we would break off, vomit over the rail of the bridge, then go down to the foredeck for another one. Even those men who normally did not become seasick were affected. I have often noticed that heavy exercise involving the muscles of the stomach, when the ship is moving violently, is enough to cause seasickness.

When the last sheep was in place, we leant wearily against the bulkhead, soaked to the skin, and wondered what the captain would say when he discovered 24 dirty sheep straddling a foot of sodden straw on his clean bridge wings. As day dawned, I went to see how the animals were faring and was surprised at how well they had adapted. They had dried off and were standing confidently, leaning alternatively to port and starboard to accommodate to the roll of the ship and obviously quite content to spend the rest of the voyage there.

One of our tasks in 1953 at Macquarie, in addition to relieving the main station, was to replenish the supplies at a field hut at Hurd Point at the extreme southern tip of the island.

We set off from the ship in convoy, with the motor boat towing the inflated rubber pontoon. The second mate and an apprentice were in the boat, and I and 10 ANARE men sat on the stacked cases on the rubber pontoon. The captain and first mate leant over the rails of the bridge, shaking their heads dubiously and occasionally looking in at the heavy surf ashore through their binoculars.

Over the years we had devised a technique for landing stores through surf on rough shores. The essential thing was the US Army inflatable rubber pontoon, 10 metres in length and 2.5 metres in width. Macey, assisted by ANARE storeman George Smith, had devised a form of wooden decking to sit on top of the three inflated rubber tubes, and this decking was built in three end-on sections connected by hinges. In a swell or in the surf, the rubber pontoon would bend and undulate like a caterpillar and the hinging allowed the wooden decking to move also with some flexibility.

The general plan for a landing was as follows: the motor boat would tow the pontoon towards the shore until it reached a point just outside the line of the breaking surf. Here an anchor would be dropped, attached to a floating buoy, and men on the pontoon would hold onto the buoy to keep the pontoon in position. The motor boat would cast off and retreat some way further out beyond the reach of any freak breaker.

One man on the pontoon would then fire a line ashore using a rocket gun. (Some practice is needed to gain accuracy in this.) If there were men ashore they would haul on this line and drag in a heavy 'grass line' (a rope made of sisal that floats in water) attached to the end of the rocket line. When this long grass line stretched from pontoon to shore, the shore end would be firmly anchored (generally to a large boulder) while the pontoon end would be fixed to the buoy. There would thus be a permanent line stretching unbroken from buoy to beach.

The pontoon crew would then cast off from the buoy, sit in line lengthwise down the centre of the pontoon and, hauling onto the fixed rope, bring the pontoon, bow first, towards the shore. When the bows grounded, three men would spring off into the shallows and dash up the beach, one straight ahead to moor the bow rope to an anchoring boulder, and one man each to port and starboard, hauling a line tied to the stern of the pontoon. The men with these stern lines would haul on them to hold the pontoon bows-on to the beach to prevent it broaching and capsizing. In this task, they generally had to be assisted by two or three other men after they had disembarked.

If there were no men already ashore, another tricky additional exercise would be needed. A dinghy, preferably a rubber one, would be rowed in to the shore from the buoy by two men. As the dinghy was generally capsized in the surf, these men had to be prepared for a wetting. (Years later, when they became available, wetsuits greatly improved this part of the act.) They would then be available on shore to receive the rocket line.

So much for the theory. In practice, it was quite difficult to achieve, so we always rehearsed our teams beforehand, generally at Port Melbourne before the ship sailed. On this particular occasion, the second mate did not like the look of the surf and cast off the pontoon too soon. He should have held us until we had dropped our buoy and anchor. As it was, we were left drifting dangerously close to the surf line as a succession of giant swells passed under us and we worked frantically to separate lines and chains and get the anchor and buoy overboard. It was a relief to feel the anchor grip and then to hang onto the buoy to halt our drift towards the breakers.

Gordon Abbs raised the rocket gun and the men cheered as a perfect shot reached the beach at the feet of Wilkinson and his two companions. However, when they hauled in the rocket line, they found there was not enough grass line to reach the shore. The spare grass line held as a precaution should such an extension prove necessary was the line that had been hauled up onto the deck of the ship when we cast off! There we were, ready to haul ourselves in along a rope, the first 60 metres of which was thick and strong and the last twenty metres was light rocket line.

There was nothing to do but to start hauling. Each swell lifted and humped the pontoon, the decking buckled and hinged, and the stacked cases groaned and twisted as we propped to support ourselves and hauled for all we were worth to keep the pontoon moving at right angles to the surf line. The shore began to loom ominously close; the backdrop was the plateau escarpment, 240 metres high, clothed with green tussock grass on the spurs and filled with scree in the gullies. The narrow strip of pebble beach about 30 metres wide was flanked by tall pinnacles of black rock extending out into the sea as reefs and niggerheads, around which giant kelp swirled and giant waves crashed in a smother of foam and spray. Thirty metres above the beach, nestling at the foot of a giant scree slope, were the auroral observatory and living hut, the objective of our excursion.

Then, as our apprehension reached its peak, we were away and no choice remained. Picked up by a large breaker, we went creaming in toward the beach, like a surfboat at a Sydney beach. It was all we could do to move our arms fast enough to pass the rope through, and our fears turned to exhilaration as we charged in, with the powerful surge of the breaker beneath us and the wind-blown spray in our eyes.

Suddenly, everything was action. The pontoon swept up the last 20 metres in foaming triumph to ground its bows on the shingle, and the three anchor men jumped off and struggled through the water to their appropriate positions on the beach. Others quickly sprang after them as the pontoon began to drag back towards deep water in the backwash and then to charge shorewards again with the next wave, dragging sideways at the stern lines in an almost animal-like endeavour to wrench itself free.

I had positioned myself at the stern of the pontoon, perched high on a packing case, in order to take cine film of the episode. Through the viewfinder it looked really dramatic. (On two other occasions, over a space of some years, in similar circumstances, a wave bucking under the pontoon had flipped the stern up and catapulted me into the surf. Rough on cine cameras!)

It took eight men to hold the bucking pontoon, surging up on the beach and dragging back, twisting and broaching, while the men stumbled and cursed and fell in the surf and wrestled to control the monster. The remainder of the men walked into the water, grabbed cases off the decking and struggled up the beach beyond the high tide mark.

When the unloading was completed and the men had laboriously manhandled the stores up the slope and stacked them beside the huts, I took stock of our position. It was after 4 o'clock and daylight was fading. Rain had started to fall and we could see the ship only dimly, standing out to sea. The wind was rising and we were obviously in for a dirty night. We were in a difficult position. The light rocket line connecting the beach to the surf end of the floating grass line had become entangled in kelp and submerged, and no matter how we dragged and tugged we could not free it. There was no way now in which we could haul the pontoon out through the first portion of the surf. It began to look as though we might all have to walk 35 kilometres up the island to the station.

I went to the hut and called the station at Buckles Bay on the radio. I asked them to radio the *Kista Dan* and ask the captain if he had a rocket gun and line on board. If so, could he recall the motor boat, send it back with the gun, the rocket line *and* a long grass line, and repeat the exercise of firing the line ashore. A few minutes later we had his reply in the affirmative.

It was with relief that we heard the ship's siren sound the signal for the recall of the boat and, about half an hour later, the rocket came hissing ashore. It took only a few minutes to haul in one end of the grass line, then we were ready to go. The men manhauled the pontoon as far out into the water as they could and scrambled aboard, those on the stern and bow lines holding as long as possible until the men in the pontoon had it moving out along the rope. The line men then made a dash through the surf and joined the hauling team on board. The pontoon bucked a few times as it crested the breaking waves, then we were in the clear water outside the surf and ready to pick up the anchor and buoy. Night had fallen by the time we reached the ship.

By then the wind was screaming a full gale and rain was lashing the crew as they hoisted up the pontoon and motor boat onto the heavily rolling ship. With the easterly blowing, there was no chance of the ship anchoring in Buckles Bay and, as the captain did not like the look of Hasselborough Bay on the western side of the island, we simply steamed around all night in the open sea.

23
EXPLORING OATES LAND

In 1959 the ANARE succeeded in landing on unexplored Oates Land for the first time.

Oates Land, just west of the Ross Sea and extending roughly from 155°E to 160°E longitude, was one of the last areas of Antarctica to be explored. Part of its coast was seen by Lieutenant H. L. L. Pennell, RN, captain of the ship *Terra Nova* who was returning to England in February 1911 after landing Scott's British Antarctic Expedition at McMurdo Sound. To the land that he sighted between 159°E and 157.5°E he gave the name 'Oates Land' To some coastal mountains that he saw in the distance he gave the name 'Wilson Hills'.

The region is highly mountainous and the coast is rugged and inhospitable, with sheer rock bluffs and ice cliffs rising from the frozen sea. Consistently heavy pack ice extends far out into the Southern Ocean and the weather is notoriously abominable.

In 1947, airmen of the US 'Operation Highjump', operating from a seaplane tender about 200 kilometres north of the coast, carried out photographic runs over part of the region but, with no landings and no astrofix control points, little cartographic use could be made of the photographs. It was left to expeditions from the USSR (1958) and Australia (1959) to land men and carry out the first surveys.

Of all my explorations, those of Oates Land were the most rewarding and exciting.

Our 1959 attempt began quite hazardously. In the ship *Magga Dan* (Captain H. Moller Pedersen) we headed south at longitude 150°E to try to reach Cape Freshfield, but bad weather and heavy ice forced us to abandon the attempt and we turned north to reach open water.

I decided to try again further east and this time had more luck. On 17

February we turned south near longitude 160°E and in the afternoon ran into ice. In spite of fog and heavy pack we made progress throughout the next day and at 2135 hours stopped for the night only 28 kilometres (by radar) off the Wilson Hills, with what appeared to be impenetrable pack to the south.

On 19 February we moved again at 0315 hours through snow showers into close pack. Fortunately, it was one season's ice and the easterly wind was light, so we could proceed at one or two knots. At 0430 the showers cleared and we could see the coast for the first time about 25 kilometres away. It appeared to comprise rocky, snow-drifted bluffs rising sheer from the sea to a height of more than 1200 metres. Very grim and forbidding it looked. A growing mood of excitement gripped all on board as our hopes rose.

At 0530 we ran into heavy floes and were reduced to backing and charging. Snow started falling again and we could see little. On the radar we could see what appeared to be a long iceberg on our starboard side stretching for about 15 to 20 kilometres towards the coast. I considered that this obstruction had probably stiffened the pack in this area. If we had seen this earlier we would have attempted to come in on the western side of it. With a prevailing easterly wind and a westerly current it could be expected that the pack would pile up against the eastern side of this berg.

At 0730 hours the ship stopped to allow the engineer to clear ice from the water valve. When, at 0830, the captain tried to proceed again, the ship would not budge. It appeared that a falling tide had tightened up the ice, so we resigned ourselves to waiting.

At noon we tried again, but without result. By then the ice had hummocked, producing 'tombstones' all around the ship, indicating the pressure to which it was subjected. At 1430 the captain found that the ice had eased a little, so he began backing and charging a metre or two at a time, and slowly worked to turn the ship around for a retreat to the north.

At this stage the weather suddenly cleared and we had a fine view of the coast. To our consternation, we found that what had appeared to be a large iceberg to the west was, in fact, a glacier tongue emerging from a broad valley glacier amongst the Wilson Hills and extending about 30 kilometres out to sea. We discovered by radar that we were gradually drifting towards this feature, which was then little more than 1.5

kilometres away. It was essential that we proceed north if we were not to be crushed against the ice cliffs of this barrier by the westerly drift.

The atmosphere on the ship was tense. Everyone was aware of the danger.

I sketched and photographed the coast but had difficulty in reconciling it with the description which Pennell gave of it in 1911. I felt sure he did not see the glacier tongue or the features to the west of it. I therefore decided to recommend restricting the name Wilson Hills to the mountains eastward of the glacier tongue. Pennell himself had no feature named after him, so I called the tongue the Pennell Glacier Tongue and the glacier giving rise to it the Pennell Glacier.

Our progress was painfully slow and we continued to drift towards the tongue. Gradually, as the ship cleared more space and could make a longer charge, we improved our progress, making about half a ship's length each charge. By 1700 hours, by working the ship hard, the captain was just able to clear the tip of the tongue, which by this time was little more than 300 metres to port. It was a great relief to be free. The strong westward set of the current was apparent, for to proceed north we had to steer about 025° true.

At 1800 hours things improved as we moved out of the tombstone ice and by 1900 we were able to proceed slowly without stopping. At 2000 we could see an open pool about two kilometres ahead.

I anticipated that there would be open water, or at least loose pack, in the lee to the west of the glacier tongue Beyond the tongue in that direction we could see the coast swinging west-north-west with two mountain outcrops not far inland. It was apparent that there was much valuable work to be done if we could have some luck with ice and a few days of fine weather. At 2200 hours we stopped for the night in ice which had thickened up again, about two kilometres short of the pool.

Friday 20th dawned grey and overcast with a south-west wind and the barometer steady at 974 millibars. The ship started at 0430 hours and, whether because of the tide or the different direction of the wind, found the ice easier to handle. By 0500 we were in fairly open water and headed westwards. At 0645 we reached an extensive pool about a mile across and I decided to carry out a flight. We appeared to be about 90 kilometres from the coast and visibility was good. If our luck turned bad we might not have another chance to photograph the land.

Doug Leckie took me up in the Auster at 0845 hours and our take-off was exciting. We were 150 kilograms overweight and the underpowered Auster barely managed to scramble off the water and limp into the air before reaching brash ice at the end of our pool. We then set off on a course of 210° for the most westerly of the mountains, which we later found the Russians had called Gory L'va Berga. West of these mountains was a low area (a bay filled with ice shelf or bay ice) out of which rose a steeply domed ice island. Beyond this the coast swung away to the north-west. Between the Berg Mountains (Gory L'va Berga) and the sea the ice was low and broken up, giving the appearance of ice shelf or very thick bay ice, and this extended eastwards past several nunataks. Beyond these to the east were mountains, then the Pennell Glacier and Tongue, then the rugged rock bluffs of the Wilson Hills. Fast ice adhered to the coast as far east as the nunataks, but a polynya extended from there to the glacier tongue.

From the Berg Mountains we flew on course 115° true at 600 metres. I carried an F.24 aerial camera and took a series of hand-held overlapping oblique photographs, one about every 20 seconds, until we had crossed the glacier tongue and reached the Wilson Hills.

It was an uncomfortable flight. My feet were cold and I was very cramped carrying the heavy camera. I had to hold it half out of the small window of the plane and the hand triggering the shutter became frozen. Just off Williamson Head, Leckie said fuel would not permit us to proceed further, so we turned for the ship—or for where we thought the ship should be. Our return made the third side of an equilateral triangle, each side being roughly 90 kilometres.

The extensive pool in which we had left the *Magga Dan* had been surrounded by pack-ice and icebergs. With the large camera on my lap I had been unable at the beginning of the flight to peer around as I usually do in such circumstances and commit to memory the appearance of the surrounding icebergs. Leckie, apparently, had also failed to do this, probably assuming that the pool itself could not be missed. However, drifting pack ice must have closed over the pool during our flight, for when we returned we could not find it. There was only a vast mosaic of confetti-like floes, each of the same order of magnitude as the *Magga*. Our radio compass would not give any indication, but we were in radio communication with the ship.

I was not the only Antarctic explorer to have experienced this difficulty. The Norwegian explorer-aviator, Hjalmar Riiser-Larsen, after a flight made on 7 December 1929 off the coast of Enderby Land, wrote:

> On this flight we experienced how immensely difficult it is to find a ship in the ice. What you look down upon is not a huge white plain; the ice is broken, some cakes drift loosely, some are closely packed; the picture, indeed, in black and white and in colours from the diatoms, is the most irregular that could be dreamed of. Moreover, if one is away any length of time, ice conditions may have completely altered before one gets back.

By radio I requested the ship to make smoke; but a diesel engine cannot make smoke like a coal-fired steam engine and we could see nothing.

I then asked the ship to take a bearing on one of the mountains on the shore. We flew halfway back, lined up the mountain on the reverse of this bearing and headed back. Still we could see no ship.

By this time our fuel was getting low. Our situation was parlous in the extreme. There were no open water pools and no large floes. We had no way of landing the Auster, fitted with floats, on the available floes. And even if we survived a rough landing, how would the ship find us? It would not know whether to go north, south, east or west; and if we could not see the ship from the air, it could not possibly see us at sea level.

I tried one last stratagem. Radioing the ship, I asked the captain to send men up on to 'monkey island' (the flat, open roof of the bridge) with every available pair of binoculars, to divide the sky into sectors, to give each man a sector to scan and to see if anyone could pick us up. One man spotted us as a speck in the sky and the radio 'talked' us back to the ship.

But our troubles were not over. Where were we to land? The captain now took the initiative; he signalled full speed ahead to the engine room. In the fairly close pack the ship could not move much, but the turbulent wash from the propeller washed away the ice immediately astern, producing a lane of open water about 40 metres long.

I was lucky to have a pilot as experienced and competent as Leckie. He had had wide experience in landing on restricted airfields in New Guinea and, later, in the Australian Alps while working for the Snowy Mountains Authority. He flew straight at the stern of the ship and, to me, it appeared that we were to crash onto the after deck. Then, at the last moment, he dropped the floats into the water at the beginning of the water lane and the heavy drag of the floats brought us to rest with our propeller almost touching the stern. We had less than five minutes of fuel left.

Next day we had a great run and reached the coast to make our first landing in Oates Land, just west of the Pennell Glacier Tongue. I raised the Australian flag on shore and Burnside and Hollingshead obtained a good astrofix. Several of our men climbed a nearby mountain and placed a message beneath a cairn they built on the summit. I later named the mountain Magga Peak after our stout ship.

On 21 February we rose at 0330 hours and Captain Pedersen moved the ship to a point where the ice edge was strong enough to permit the unloading of our Auster aircraft. Overnight the RAAF team had converted its landing gear from floats to skis, so Leckie went ashore to lay out markers delineating the best ice for an airstrip.

I wished to view the coast westwards from the Pennell Glacier Tongue as far as Cape Freshfield if possible.

It was 0640 hours when we departed, the Auster rising smoothly off the excellent surface. The day was calm and overcast and we could not fly higher than 610 metres because of cloud. We started on course 295° true at 60 knots but 15 minutes later changed to 310°. From a nunatak just west of the ship, a jumble of bay ice and ice shelf ran into a broad bay which swept westward in front of several nunataks. The ice shelf continued on, extending half-way 'inland' towards the Berg Mountains. In front of these the ice was riven by a series of giant cracks—crevasses as wide as boulevards which, nearest the mountain, lay at about 20° to the direction of the coast and which swung to lie at about 40°-50° to the coast as they reached it. There, the long parallel fingers of ice between the cracks protruded out into the sea, forming a glacier tongue which exhibited longitudinal valleys and ridges on its surface similar to those on the giant iceberg Leckie and I had examined on the Virik Bank on 12 February. The acute angle at which this tongue projected into the sea was unusual. A number of the fingers began breaking up into block iceberg form about halfway between the mountain and the coast.

To the west of the Berg Mountains, the ice shelf continued and its southern limit passed behind a high, domed ice island about 25 kilometres long which rose from the shelf and in front of a small, almost snow-obscured, ridge-shaped nunatak about 20 kilometres in from the coast. This was the last visible rock outcrop. The snow slopes of the mainland then seemed to swing north again to where a long iceberg tongue jutted into the sea. The coast was cluttered with a fantastic jumble of icebergs, bay ice and tumbled pack, and ran in about direction 305° true.

At 0710 hours we passed by the tip of the ridged iceberg tongue and flew over a comparatively small bay choked with bergs and heavy bay ice. The bergs debouched from a great mass of seracs and crevassed ice on the plateau—a wicked looking graveyard of giant tombstones. Neither Leckie nor I felt very comfortable in our tiny single-engined plane, for there was absolutely no place where we could land if the engine failed.

At 0720 the coast swung still further northwards and we followed it on a true course of 335°. We had now passed beyond radio range of the ship and the plateau and bay still offered no possible landing areas. Miles out to sea the convulsed masses of heavy ice petered out into fast ice which extended a further 100 kilometres to the north.

We reached what looked like the end of the northward-trending coast at 0745 but found beyond it a small bay and a further headland, bearing 325°, and we changed course to follow it. I was most anxious to reach this point and see what lay beyond.

At 0750 Leckie pointed under our port wing and, to my surprise, I saw on the horizon some long, low, flat-topped bluffs of black rock which bore a singular resemblance to Horn Bluff. (I had flown over Horn Bluff, lying to the west of Cape Freshfield in George V Land, the previous year.) I could not believe we were close enough to Horn Bluff for it to be visible from our altitude. Under the starboard wing, and much closer, was another set of nunataks which appeared to lie on the far side of the northward-trending promontory we had been following. I asked Leckie to point the aircraft at each of these two major features in turn and found that the Horn Bluff features lay on bearing 245° and the nunataks on bearing 275°. I took kodachrome photographs from the window. (The features resembling Horn Bluff are those which we later named SCAR Bluffs.)

At this stage Leckie suggested that we should return. He was obviously uneasy. I attributed this to the desperate nature of the terrain over which we were flying and the fact that we were out of radio range of the ship. So I urged him to continue, pointing out that I particularly wanted to turn the corner of the promontory ahead and see what relation the nunataks bore to it. I promised that we would turn back after looking around the corner.

However, a few minutes later Leckie said, 'The engine's running rough. We had better go back.'

He was right. So there, 120 kilometres from the ship at 0725 hours, we turned about.

Leckie varied the carburettor heating and other controls and was very worried. Suspecting carburettor icing, he dropped down, first to 365 metres and then to 240 metres. This terrified me, for at that altitude we would just go straight in if the engine failed. With our ears straining to the rough noise of the engine and our eyes searching the terrain below for non-existent landing sites, we skimmed over the horrible bay of cluttered icebergs, over stretches of crevassed plateau ice and over 'tombstone' bay ice. Meanwhile, a new anxiety was added to our minds—the weather began to close in from the east. Low dark clouds hung in the sky and a pall of mist was slowly enveloping the mountains around the Wilson Hills.

Leckie twiddled again with various controls and the engine smoothed out. We found out later that a petrol pump had developed a fault. Thinking it was a carburettor icing, Leckie had increased the carburettor heating. At about the same time we had changed to another tank, as the other was running low in fuel. This change automatically brought another fuel pump into action, thus removing the original source of trouble, but the extra carburettor heating continued to cause the engine to run rough until, having decided that icing was not the trouble, he restored the carburettor heat to the normal level and the engine thereafter ran satisfactorily. We, however, were not to know all this and continued to worry all the way back.

We arrived over the ship at 0845 and, as the ceiling had not closed in completely, I asked Leckie to fly on over the Pennell Glacier Tongue to look at the condition of the pack ice on its eastern side.

We found that open water extended all around the tongue, although the bay on the other side was partly covered with medium pack extending right to the foot of the coastal mountains.

The Pennell Glacier Tongue, like the giant iceberg at Virik Bank, exhibited longitudinal valleys separated by domed ridges, like long fingers laid side by side. Several of the fingers were beginning to separate at the northern end, with sea filling the valleys between, and the ends of some of the western fingers had broken off.

This form of glacier tongue is so unusual that I could not help wondering whether the giant berg we had seen at the Virik Bank had originally formed part of this glacier tongue or the one further west which lies in front of the Berg Mountains.

The front of approaching bad weather was now quite close so we were forced to give up any further reconnaissance. We landed at 0900 hours on the fast ice and went on board for a hearty breakfast. On top of Magga Peak a cairn had been built and the Australian flag was flying. Our surveyor had obtained a good astrofix and we were content to depart.

24

MISHAP AT LEWIS LAND

On my 1956 voyage in Kista Dan *to relieve Mawson Station I decided to explore part of Wilkes Land, where US 'Operation Highjump' aerial photographs had shown several small rocky islands along a stretch of coast about 150 kilometres long. The first, in Davis Bay, they had named Lewis Islet, but no-one had ever visited it.*

On 5 January, Squadron Leader Douglas Leckie, RAAF, took off with me in a Beaver aircraft, mounted on floats, from a pool of open water amongst pack ice about 100 kilometres north of the Antarctic continent. The day was fine and sunny and we flew SSW at 300 metres with an air speed of 160 kph.

We crossed heavy floes, recently broken from fast ice and very large, completely covering the sea. Next we crossed a lead of open water, about a kilometre wide, running east and west. The other side of this there was unbroken fast ice running right in to the coast. There was no chance of the ship moving south from the pool in which she lay. We continued on, over a remarkable accumulation of icebergs—scores of them—immobilised in the fast ice and almost certainly aground. Beyond these, which extended more than 60 kilometres out from the coast, was open water leading into Davis Bay.

In later exploration we were to confirm that the bergs were, indeed, grounded and that they and the interlocking fast ice formed a permanent barrier in the ocean. We mapped the extent of this feature, whose bergs must have numbered well over 100. I named it the Dibble Iceberg Tongue. This was the first time an iceberg tongue had been recognised as a coastal feature in Antarctica.

Very close to the extreme south-western edge of the iceberg tongue, where it joined the continental ice, we found Lewis Islet, nestling practically against the coastal ice cliffs. It was rocky and bare of snow and it looked as though we would be able to land on it. Pushing on, we measured that, inland, the continental ice sheet rose fairly rapidly to an altitude of 760 metres in the first 10-12 kilometres. No mountains were to be seen inland. I made a rough sketch of Davis Bay, the ice coast and the iceberg tongue.

The coast here was quite unusual. West of Lewis Islet there was the normal pattern of the vast continental ice sheet sloping down from the plateau inland to terminate in ice cliffs 15-30 metres in height. In places the ice sheet would push on past the continental rock margin which supported it and become a floating platform. From this platform, from time to time, flat-topped icebergs would break off and float out to sea.

However, just east of Lewis Island, the pattern was quite different, providing a phenomenon I had never before observed. Instead of featureless plateau ice, the surface near the coast appeared as numerous interconnected 'mounds', each about 400 metres across, with heavily crevassed tops. It was as though one had made a slab of pastry and dropped it onto a bed of marbles. These mounds were separated at the coastline with the spaces between filled up with accumulated snow. Further north, the separations increased until, out to sea, in the middle of the tongue, the mounds became distinct icebergs separated by frozen sea (fast ice). At the western edge and at the northern tip of the tongue, the icebergs were free-standing in open water. It is probable that the whole system, over many decades, gradually moves northwards until the shoal bank upon which it rests dips to greater depth and the icebergs float away in deep water.

On the return trip we noted that the pack ice to the west of the iceberg tongue was much more open and that the ship would be able to push through. Obviously the prevailing easterly winds had packed up the ice to the east of the tongue where, over winter, it had frozen solid.

On Friday 6 January the *Kista Dan* moved west, around the tip of the iceberg tongue, then south into Davis Bay. On the way we passed a giant berg more than 32 kilometres long. We had difficulty finding Lewis Islet. What from the air looked so simple, on the surface was most confusing. Davis Bay was much broader than I had thought from the reconnaissance and the ice cliffs at the edge of the continent stretched uniformly and unbroken to both east and west. Finally, at 0100 hours on Saturday 7th, we anchored about half a kilometre north-west of the islet and went to bed.

At 0400 hours I rose and looked at the weather. It did not promise well. There was a south-east wind of about 20 knots and a lot of low cloud, so I ordered breakfast for 0700.

At 0830 the ship's motor boat, driven by Dick Thompson, landed Bill Bewsher (the OIC designate for Mawson Station) and me on a ledge of rock at the northern tip on the island, then returned to the ship to bring the team for a program of observations on the island—surveying, geomagnetism, gravity measurements and biological investigations. First came Kirkby (surveyor) and McGregor (geophysicist) with their instruments. The following trip brought Ayres, Bunt and Dowie. While they were coming, Kirkby and McGregor selected a site for their observations and we carried their gear up to the top of the island.

As the day wore on the wind decreased and the aircraft was able, in the afternoon, to carry out photographic runs of roughly 160 kilometres both east and west of Lewis Islet. Ashore the work went well and the motor boat made a number of ferry trips back and forth carrying expedition men who wished to look at the islet. Bunt (biologist) and Dowie (medical officer) began their ornithological survey.

The islet was exceptionally interesting. It nestled closely alongside the ice cliffs of the continent, separated from them by a mere 30-35 metres of sea. It was about 270 metres long and 180 metres wide, of roughly oval shape, and its rocky surface rose in the centre to a height of 27 metres. As the only rock outcrop for more than 160 kilometres of coast, the island was a breeding ground for great numbers of birds. Our biologists made a census of the Adélie penguins, which numbered over 5000.

There were, in addition, many nesting silver grey petrels, cape pigeons, Wilson's storm petrels, Antarctic petrels and skuas.

The census was taken by a number of men moving slowly forward in line across the island and counting the number of nests and the numbers of adults and chicks in them.

The Adélie penguin lays two eggs but, mostly, only one chick survives. The birds were nesting in 'groups', each compromising between 10 and 40 pairs. The locations of the groups seemed to be determined by the areas of rock that first became snow-free when the birds arrived from the sea to nest. We had arrived in the middle of the hatching period, for some nests contained eggs while others had chicks. Less than 30 per cent of nests had two chicks.

I saw one fair-sized downy chick wandering around one group, apparently separated from its parents. Each time it tried to attach itself to an adult the adult would peck it and drive it away, even though in some cases the adult appeared to have no chicks of its own.

I watched in amazement as the Adélies landed on the islet. Around most of the islet's edge a fringe of ice covered the rocks, the result of frozen spray and, in some cases, snow drifts. The penguins would propel themselves out of the sea to land on a slab of flat rock, sometimes landing vertically on their feet and sometimes horizontally on their bellies. They would then use their claws as crampons and simply walk straight up ice sloping at 60° or 70°.

The silver grey petrels were the men's favourite birds. About the size of a domestic hen, they are delicately coloured in white and light grey with soft, downy, thick plumage. They nest in open nests on flat ledges of rock on cliff faces, generally with a vertical rock wall behind them and facing the sun. They are so tame that, if you approach quietly, you can gently stroke them—or even pick them up to examine their eggs.

Our surveyor, Syd Kirkby, obtained an astrofix to determine accurately the position of the island, which worked out to be 66°06'16"S latitude, 134°22'17"E longitude. This localisation of the solitary island was essential for tying down the exact positioning of the aerial photographs obtained by the pilots.

In preparation for New Zealand's Antarctic program for the International Geophysical Year (1957-58), an alpine guide from Mount Cook, Harry Ayres, had been sent by Edmund Hillary as a New Zealand observer with our expedition. Hillary and Ayres had done much ice climbing in New Zealand but neither had before that time visited Antarctica.

Bill Bewsher was a competent mountaineer with a lot of experience of rock climbing, but he had never performed on ice. He looked at the ice cliffs across the narrow gap of water separating the islet from the mainland and said, 'Phil, why don't we go over to those cliffs and get Harry to give us some practice in ice climbing.' It seemed a good opportunity, so I called Harry and, together, we went down to the landing place, summoned the motor boat and collected the necessary gear—climbing ropes, ice axes and crampons.

There were only two ice axes. To obtain another would have meant sending the boat out to the ship again, a delay of at least half an hour. Very foolishly, I said, 'We won't waste time. You and Harry can use the ice axes. I'll be all right without one.'

The motor boat landed us at 1400 hours on a small ledge of rock at the bottom of a 20-metre ice cliff, then went back to the ship. The men on shore were on the far side of the island from us and were not visible from where we were. We donned the crampons, roped up with Harry leading and me in the middle, and began to climb.

Harry led us on a long traverse across a slope of about 60°, cutting large steps 'buckets' he called them) for us to step in. When Harry was close to the top, on a 70° slope, I was on a steep pinch about six metres below him, while Bewsher was on a milder slope further down. Below him was a vertical ice face that fell 10 metres into the sea.

I heard Bewsher call and, looking around, saw him sliding down the slope. I took a turn of the rope around my shoulders and squatted down in my 'bucket' to belay him, but when the rope tightened, having no ice axe to hold on to, I was pulled from my stance and began tumbling downwards.

I remember being quite cool and calculating. I was thinking—Bewsher is gone, now I'm gone, Harry won't know and, when the stress comes, he will be torn from his foothold and the three of us will fall into the sea. There will be no-one within sight or earshot, there is no place below us where we can scramble out, so we shall probably thrash around in the sea for five minutes or so before we freeze and drown.

Of course I should have shouted, but I was too busy thinking. Luckily Bill saw me spill out and gave a great yell. Harry dug his ice axe into the cliff without even looking around, took a turn of rope around it and, lying on it, withstood the tug of the rope and held us both. He then carefully brought us up, one at a time, belaying us on the rope until we were back in the 'buckets' behind him. It was not far, then, to the top of the cliff. On Mount Cook, New Zealand, he had probably done this many times with novice climbers.

Crampons are awkward things for the beginner to use. Strapped onto his feet, the three-centimetre-long spikes extend the length of his leg so that, on each step, he has to lift his leg higher to have the spikes clear the ground. On a steep slope, the inside leg has to be lifted very high through the narrow gap between the outside leg and the slope.

In Bewsher's case, he had obviously caught his inside crampon spikes against either the ice surface or his outside trouser leg and had tripped himself.

Standing on top of the ice cliff, still roped together, we decided to walk some distance inland, up the plateau slope, hoping to see over what looked like a crest a few hundred metres ahead. But as we walked, the apparent crest just kept receding, so we gave up and returned.

But on the way up, there had been another incident. Bewsher came up to me, laughing and chatting, obviously wanting to celebrate his joy at our lucky escape. I screamed at him—'Go away! Don't come near me! Stay at least, six metres away and make sure the rope joining you and me is at an angle to that joining me to Harry! This is crevasse country and we must never walk in line or we might all go down together.'

Less than two minutes after this, the snow gave way beneath me. I instinctively threw out my arms and finished up supported by them, flat on the snow at shoulder height, spanning a gaping hole in which my body hung. Looking down I could see the blue, shimmering, icy walls of the crevasse disappearing into the black depths.

Bewsher and Harry tightened their ropes and helped me to claw my way out.

The view was magnificent! The ice-edge cliffs, the bay full of icebergs, the blue sea, the warm sun and the calm, placid day all combined to make an occasion of perfection. From the cliffs, periodically, would come a roar as of a blasting explosion and tons of ice from the edge would crash down into the sea. Inland, similar explosive sounds told of the ice cracking open. Again, I saw the phenomenon of irregular expansion under the action of a hot day's sunshine cracking off the surface layers of the ice cliffs.

The motor boat duly arrived back at the foot of the cliff at the appointed time and took us back to the ship. After dinner, I worked at my desk, writing reports, until midnight, then had a shower and went to be at 0045 hours.

It had been quite a day!

25

HELICOPTER CRASH, VINCENNES BAY

The first, and only, serious aircraft crash in ANARE operations in my time occurred south of the present Casey Station in 1960.

On Saturday 13 February 1960, I was directing the third day of our relief of the ANARE Station at Wilkes. It was fine, sunny and calm when I rose at 0530 hours in my cabin on *Magga Dan* so I arranged for the two helicopters to fly David Cook (surveyor) and Ian McLeod (geologist) roughly 96 kilometres south, to the bottom of Vincennes Bay, to take observations from outcrops of rock—the Hatch Islands—nestling against the edge of the continental ice sheet. The pilots were Ray Hudson and Peter Ivanoff. They got away at 0815 hours and I remained on the ship supervising the unloading of cargo.

At 1000 hours the ship's radio officer brought me an urgent message—Ivanoff's helicopter had crashed on the mainland ice. He was injured but David Cook, his passenger, was all right. Hudson had landed his helicopter safely on the Hatch Islands' rock. His passenger, Ian McLeod, was attempting to climb the plateau ice to reach the crashed aircraft. David Cook was erecting a tent beside the smashed helicopter.

This sudden emergency set in train a series of activities. The ship had to prepare to leave her anchorage. I had to call all men aboard from on shore, explain the circumstances, select those I needed, and send the rest back to continue work. With several others I discussed what we needed to take for a rescue operation—sledges, climbing ropes, ice axes, skis, crampons, pitons, carabiners, petrol stoves and fuel, thermos flasks with hot soup, sleeping bags, rations, splints, plaster of paris, various drugs and medical instruments, etc. My lieutenant, Bob Dalton, was given the job of supervising the collection of all these items. I at first decided to take both doctors but, just as the ship was ready to leave, I received a radio message from Hudson's helicopter saying Ivanoff was not immobilised so, as the OIC on shore (Bob Dingle) was reluctant to be left without a doctor, I sent one of them ashore. We sailed at 1120 hours and had a fast run down Vincennes Bay, luckily being able to pass through a gap in a belt of heavy pack ice encroaching from the west. The trip took about six hours. As we approached the bottom of the bay, the wind increased until, when we anchored half a mile off the coast, it was around 40 knots. Obviously there was a high katabatic content in this, with a gravity wind rolling down the steep coastal slope from the plateau.

Our motor launch was hoisted out and I boarded it with Dalton, Dr Soucek, Stinear, Clayton and White. I took the wheel and we butted our way in against the 40-knot wind and a very rough chop, being whipped by freezing spray each time the bows bashed the waves. Over the last 100 metres I handed over to Dalton so I could photograph the islands, the ice cliffs and the smashed helicopter perched on the ice just above them.

Dalton took the approach far too fast for what was a completely unknown stretch of coastal water and ran us aground some 30 yards out from the most westerly island. I took the wheel again, managed to unstick the launch, and cautiously edged it in to a landing point on the rock.

We were all very relieved to find the four men well and in good spirits. Ivanoff had cuts to the face and Cook a gash across his ear and both had their heads bandaged. Their parkas were spattered with blood and they certainly looked the part of 'walking wounded', but they were laughing and joking. I sent the launch back to the ship with all except Hudson, Stinear, McLeod and myself. Stinear and McLeod set off on crampons for the wrecked aircraft, hauling a light sledge. As the wind had dropped considerably, Hudson boarded his helicopter and flew it back to the ship. I watched to see him take off safely, then donned crampons and, carrying an ice axe and rucksack, followed the other two up onto the plateau. The surface was hard, blue, slippery, sloping ice and the distance was about 800 metres.

We found the helicopter to be a complete wreck, only the engine and dashboard instruments appearing to be worth salvaging. We could not find Cook's survey chronometer, so I roped up and moved cautiously to the edge of the cliff, but saw no sign of it or any other gear.

We collected the other survey instruments, the survival equipment (including tent) and such aircraft instruments as we could salvage. Loading the heavy items onto the sledge and the lighter, more delicate instruments into our rucksacks, we plodded back across the slope and down to our landing place, with ropes attached to the sledge fore and aft to stop it side-slipping. I had thought of trying to salvage the engine, but it would have meant lowering it down from the top of the ice cliff into a boat below, and the danger of part of the cliff ice crashing down was too great.

The launch duly returned us to the ship and we sailed for Wilkes at 2035 hours. We had to sail slowly during the dark hours and some pack ice further delayed us next morning. We anchored at Wilkes at 0545 hours on 14 February.

The airmen provided me with further information, as below:

Hudson:
We refuelled satisfactorily on the flight south at the Vanderford Glacier in winds of 35 knots. At the bottom of Vincennes Bay we first tried to land on the eastern islands of the Hatch group but the winds were fierce. Snow was blowing over the rocks and the water was whipped into foam in the sea offshore. Sheets of water about 20 yards by five yards were being torn from the surface and hurled through the air. The wind was over 60 knots and the helicopters were 'crabbing' sideways. I advised Peter over the intercom that we would abandon these islands and look further west. We then saw the island upon which I later landed and descended to have a look at it.

Then everything happened at once. I found myself flying higher than Peter. I was flying at full speed, with my airspeed indicator reading 80 knots, yet I was making no forward progress. I saw Peter's chopper flapping up and down in the turbulence before I got caught in it. With full climbing power I was rushing downwards. I made a tight turn and aimed to land on the island. I barely succeeded and, as I landed, I looked up to see what Peter was doing. With a shock I saw his helicopter crash and bits of aircraft fly into the air or slide down over the cliff below into the sea. I turned away, expecting the plane to follow the pieces over the cliff. When I did look up, the chopper was still there and I saw David climb out and walk around, but not Peter. I assumed he was injured.

Later I saw Peter apparently lying on the ground near the plane. He waved and I thought, 'He is conscious but has injured his back or a leg'. (As it happened he was all right but simply couldn't stand up on the slippery ice.)

My passenger, Ian McLeod, set off on crampons from the island, cutting steps with his ice axe on the steep, lower slopes of the plateau on a long traverse to climb to the helicopter. I stayed in my helicopter for the first hour, talking by radio to the ship. Then I unpacked our survival gear, carried it down to the small ice valley between the island and the mainland, pitched the tent and brewed some hot drink.

Ivanoff:

I was caught above the plateau, at about 1000 feet, in a severe downdraught. I had only three choices. One was to try to fight my way forward and upward over the plateau; this was not possible—I was going rapidly down even with full climbing power. Or I could peel off away from the plateau and dive for the sea. The trouble with this was that I would finish up skimming the water and there was a fifty-fifty chance I would go in, in which case David and I would be finished. So I decided on the third possibility and tried to land on the plateau. With zero forward speed and full climbing power I struck the ice with a downward speed of about 20 knots on the right-hand side. This smashed up the landing gear, the fuselage, the rotors and the perspex cabin but, fortunately, drove various pieces of the undercarriage hard into the ice, which anchored the plane and stopped it from sliding over the edge, from which we were only about 30 metres.

I was cut over the eyes probably from striking the astro compass, or even the broken perspex dome. David had a gashed ear. We staunched the flow of blood from our wounds, then roped up and headed on the long dangerous traverse down towards the rock island. David had crampons but I did not and could not stand on the slippery ice, so proceeded partly by crawling and partly by David cutting steps with his ice axe. The wind was about 50 knots, which did not make it any easier and, after some time, fatigue and shock gave us the shakes and we had to stop and rest. Eventually we met Ian McLeod who helped us down the rest of the way. The 800-metre journey took us three hours!

Ray was waiting for us in the tent with hot drinks and bandages.

When I returned to the ship, shortly after 2000 hours on the 13th, I drafted a series of radiograms for Australia—our office, the department in Canberra and the press—then interviewed Ivanoff, Cook and Hudson on a tape-recorder. I then wrote up my diary and sorted and catalogued the films, still and cine, that I had taken. I got to bed at midnight, very, very tired.

26

THE LAST VOYAGE

My two most dangerous ANARE voyages to Antarctica were my first (1954) and my last (1966). In each case hurricane force winds created the hazards.

In the late afternoon of Saturday 12 February 1966, I was on board MV *Nella Dan* (Captain Wenzel Gommesen) moored in Mawson Harbour. It was to have been our last day at Mawson; the traditional change-over party was to have been held that night and we had planned to depart next morning for Davis Station.

At 1915 hours I collected several of my colleagues (Macklin, Woinarski, Afflick), together with the captain and the chief engineer (Hugo Larsen), and summoned a DUKW to take us ashore where Morrison, the 1966 Mawson OIC, was to provide us with pre-dinner drinks.

Just as we were gathering on deck, the wind freshened, drift snow began to blow and within minutes it was gusting at over 50 knots. Captain Gommesen asked to be excused, saying that in view of the deteriorating weather, he and the chief engineer had better remain on the ship. The rest of us went ashore then sent the DUKW back to pick up our 1965 Mawson men, together with 14 of the officers and crew of *Nella Dan*. By then it was blowing steadily at over 50 knots, with heavy drift, and conditions were at the limit for DUKW operations.

On shore the party got away to a fine start with Towson playing good jazz on the piano and Williams backing him on a full set of drums he had brought south. The cook, Ralph Dyer, had prepared a grand feast, beginning with prawns, smoked oysters and lobster, followed by hot savouries and then various casserole dishes with rice. For a sweet he had baked a huge cake, decorated with coloured icing sugar depicting Antarctica, divided into national claim segments, with the ANARE leopard seal badge in the centre and blue ocean all around.

Accompanied by white wine, red wine and beer, and followed up with port, the dinner was a memorable one.

At 2230 hours I went outside to check on the weather, for it would soon be time to send the officers and sailors back to the ship. It was blowing around 60 knots and heavy drift snow was limiting visibility. Fortunately I had come ashore wearing a down-filled parka and was cosily warm from the waist up but, as I was wearing ordinary trousers and no windproofs, my legs became very cold. Through the swirling clouds of drift I could dimly see an occasional flicker of the ship's lights. Going to the radio hut, I spoke to the captain and suggested attempting to return his men to the ship, but he thought we had better not risk it, so I returned to the party and advised the men.

By midnight the wind speed was over 70 knots and the ship was quite obscured. Huge snowdrifts were forming around the huts and moving around outside had become very difficult—even dangerous for the inexperienced men—so various receptacles were improvised in the cold porch of the mess hut to act as piss-pots. Altogether there were 78 men ashore. There was no way we could collect mattresses and blankets, so eight of the 1965 party battled their way to the old recreation hut, where they knew there was some bedding, while four others managed to reach one of the caravans. The rest partied on, with music and a showing of cine films, before lying down on the floor or on chairs and trying to sleep.

I stuck it out until about 0230 when I decided to try to reach the old recreation hut, but by then it was blowing at 80 knots and I couldn't make it. Large drifts had formed to block the well-known tracks between the huts, the bare rock had iced up so that one couldn't stand on it and the night was pitch dark. The wind battered me as I staggered around and I was blinded by the blowing snow. I therefore took a shorter route to the OIC's hut, where Ken Morrison was already in bed and asleep.

I lay down on the linoleum floor and tried to sleep but my legs were cold. I turned up the oil heater and spread an old suit of overalls over my legs and had just begun to doze off when the telephone rang. It was 0330 hours.

The message from the Mawson radio officer informed me that the telephone line to the ship had broken and that the captain had contacted him by radio to say that lines mooring the ship had parted. As the ship had drifted towards the rocky edge of Mawson Harbour the captain had used the engines to push the ship forward against the wind and to wedge her bows into a rim of shore ice. There he held her in position, bows into the wind, with engines at full speed. Although the forward part of the hull was resting on the rocky bottom of the harbour, the captain had said that the ship was in no danger and that there was nothing that we, on shore, could do.

I gave up the idea of further sleep and decided to see how the men were faring in the mess hut. So I donned my parka and pushed out through a porch half-filled with snow to find a great drift almost blocking the exit. I stumbled in the dark across to another snowdrift behind the next hut, crawled over it half-smothered in snow, and tried to walk up a stretch of exposed rock to the door of the mess. I was blown over twice and then, when almost at the door, my boots slipped on the ice-covered rock and I was hurled back in a long glissade for 10 metres. So I got down on all fours and crawled up the slope, into the wind, towards the mess hut.

The wind strength was now 85 knots, with gusts to 100 knots, the strongest that I have ever had to proceed against. My face and hands were bitterly cold and the confusion of mind, produced by the stinging, blinding snow, the buffeting of the hood of my parka against my ears and forehead, the roar of the wind and the difficulty of peering out around the funnelled parka hood is impossible to describe.

I could not help thinking what it must be like to have no shelter or to be lost under such conditions. An effort of will is required in order to think at all and to resist panicking.

I reached the door but, with all my strength, I could open it no more than about 10 centimetres against the wind. After several attempts I got my hand inside, then my arm, then my shoulder and, risking what seemed probable decapitation, my head and the rest of my body.

My parka was iced up and the hood and pockets filled with snow. As the temperature was only at 5°C below freezing point, it was not cold enough to keep the snow powdery, so my gloves and trousers were caked with wet snow.

I went to the kitchen, stepping carefully over the prostrate men on the floor of the mess, and yarned to a couple of men while I brewed hot coffee. All around was confusion. Bodies lay on every spare bit of floor amongst the litter of the party—beer cans, plates of food remnants, ash trays and magazines. In places men had lined up several chairs to form beds while, in the cine projection cubicle, a member of the crew had vomited over the mailbags that he had spread to act as blankets.

At 0630 hours I breakfasted on tinned fruit and then, with the cooks preparing boiled eggs for the troops, I wandered up to the music room, found a loose settee seat lying on the floor, so lay down on it and fell fast asleep.

When I rose, at 1000 hours, I rugged up again and staggered across another wind-swept, drift-hammocked slope to the radio room, where I contacted the ship by radio and spoke to the captain. He considered it most fortunate that the storm had struck before he had gone ashore. He might otherwise not have been able to return. As it was, he had been able to take the measures necessary to prevent damage to the ship. He was still on the bridge with the engines holding the bows of the ship hard up against the ice ledge on the shore. From time to time the hull would bounce up and down on the rocks of the harbour floor.

As it was still blowing 70-80 knots there was no possibility of returning men to the ship so, with difficulty, I returned to the mess hut and organised a clean-up.

At noon we had lunch—mainly casseroles left over from the party. It was quite good, considering the circumstances and the number of men involved. Then we completed the clean-up by mopping out all the rooms. Fresh piss-buckets were installed in the porches and Woinarski and I drew up plans to provide better sleeping arrangements in case the men had to remain on shore for a second night—an allocation of floor space and two blankets for each man. Morrison had found a supply of new blankets in cases recently unloaded from the ship and stacked in an accessible spot in the open. They had not drifted over too badly.

As the men settled down to read and listen to the radiogram I rugged up again and pushed across to the distant meteorological hut where I examined the record of the night's blow. The anemometer had recorded gusts over 100 knots and an average wind speed, between 0300 and 0600 hours, of 86 knots.

Just after 1500 hours, the drift eased and we were able to see the ship dimly for the first time. It was quite frightening to see this large vessel ashore with her bows protruding over the ice and rock.

By 1600 hours the wind had dropped to 45 knots and the barometer was rising. It was time to make an attempt to evacuate the visitors.

Hawkings and I examined a DUKW. The cabin was filled with snow, so we dug it out. He then managed to start the engine and, at 1730 hours, he drove the first load of men to the side of the ship—not a long trip!

The wind was 35 knots and there was not much drift snow. The second load followed shortly after. It was a very relieved group of officers and crew who settled into their familiar and comfortable quarters.

The captain told me that, as the night wore on, he had been anxiously watching the wind increase and, as a precaution, had kept the ship's main engines running. Suddenly, at 0325 hours, two steel bow lines had parted and the third had snapped the length of railway line to which it was anchored. The ship had begun to drift rapidly astern towards the rocky West Arm of the harbour. Visibility was zero—even the foredeck was invisible because of dense, driving snow. Revving up the engines to full speed ahead, the captain had used the radar to guide him as he deliberately ran the bows aground, with the stem against a ledge of shore ice near the dog lines. To allow this he had been forced to jettison one of his stern lines. With full power he had barely managed to hold his position; in fact, at one stage during the night, the gusts had proved stronger than the engines and had forced the ship backwards!

The immediacy of the decisions that faced the captain, in a drifting ship in pitch darkness, can be appreciated when it is realised that the lines had parted at 0325 hours and two minutes later, at 0327, the bows were aground.

The captain reported to me that mooring lines fastened to the helicopter blades to prevent them from rotating had snapped, and for quite some time the blades had been rotating at about 100 rpm before they eventually stopped—the bearings had seized up! (With no engine running, no oil was being pumped to the rotor.) So we were to have no helicopter for the rest of the voyage!

Later I was to learn of another damage inflicted by the hurricane. The Mawson flagpole, a vertical length of eight-centimetre iron water pipe, had been bent at right angles. A parked Volkswagen had been moved bodily sideways for 10 metres before finishing up against a rock ledge. A large living caravan with a sledge-type undercarriage had been blown for some metres to smash up a smaller fibreglass caravan nearby. Two aerial masts had been blown down at Hump Island. And huge drifts of snow had appeared overnight throughout the station to transform its appearance to that of mid-winter.

Next day it was too windy for the ship to leave the harbour and the captain laid out some new mooring lines. However, the day afterwards was calm and fine, so we prepared to depart. But we endured an anxious half-hour when the captain found that, with engines full speed astern, the ship refused to budge off the rocks. We had visions of being marooned there for the winter!

Alternating the engines forward and astern, swinging the rudder to port then to starboard, and winching on a mooring line laid out astern and anchored to West Arm, the captain finally persuaded the ship to slide back into deep water. Shortly after, outside Horseshoe Harbour, we were able to load the DUKWs and a few remaining personnel and then we were on our way to Davis Station.

But this was not the only exciting incident upon this, my last Antarctic voyage. Our visit to Davis passed off uneventfully and we then sailed eastwards with the object of visiting an automatic weather station that we had set up some years earlier on Lewis Island (66°06S, 134°22E), and of collecting the equipment, which was no longer operating.

Arriving on Wednesday 2 March, I found that a heavy swell breaking on the rocks prevented our landing. Next day, at 0430 hours, Captain Gommesen called me to look at the weather. The sky had become overcast, drift snow was streaming off the top of the continental ice cliffs and the wind was 30 to 35 knots. The sea was very rough and there was still a heavy surge on the rocky shore of the island. The barometer trace was plunging steeply.

As our tight schedule did not allow us enough time to ride out the storm and then work on the island, I reluctantly told the captain that I would abandon my plan and that he could depart immediately for Australia.

The ship turned north and proceeded at slow speed in 400 metres of water along the western side of the Dibble Iceberg Tongue. I spoke to the captain and said, 'I should go a bit faster if I were you. You could be caught with blowing snow and poor visibility before you clear the icebergs.' He replied, 'The men are working on the foredeck and I must not go faster because of the swell.' There was no swell and the wind and waves were astern. I thought, as he had been so anxious to leave Lewis Island, that he would have done as other captains here have always done, gone full speed ahead in the good deep water alongside the iceberg tongue.

The weather degenerated rapidly; at 1300 hours it was blowing 55-60 knots, with driving snow; by 1500 it was gusting to 73 knots and by 2000 it was 85 knots—hurricane strength.

By the time night fell we had cleared the northern extremity of the iceberg tongue but were not out of the pack ice zone. We were heading into the easterly storm and making a few knots. The ship was pitching and rolling abominably and I was closer to being seasick than I had been for years. I lay down in my clothes on my bunk—it was uncomfortable and exhausting trying to stand or walk around.

The captain spent a dreadful night on the bridge, standing by the engine controls, peering through the clear-view rotating glass screen, seeing nothing except the sea immediately to port and starboard of No. 2 hatch, with an officer at the radar and men on lookout on the bridge wings. At 0200 hours on Friday 4 March, the ship made several exaggerated rolls—more than 50° from the vertical—and I got up to re-stow a number of articles that had come adrift in my cabin. I then went up to the bridge to see what was happening. I found that we had run into brash ice—pulverised fragments of ice floes smashed up by the waves—and so the captain had turned the ship around to run with the wind in a westerly direction in the hope of reaching ice-clear water. The heavy rolling had occurred as we turned about.

It is always disconcerting to have to turn a ship about in a high wind, for there are those few moments when, abeam to the wind, the ship must take the waves broadside; but in a hurricane the waves are breaking and the force of the wind adds to the thrust that heels the ship over. The danger of capsizing becomes very real.

With a ship as stable as the *Nella Dan* the captain had decided that capsizing would be less of a risk than being battered by heavy brash ice.

Icebergs were not the main danger. They showed up clearly on the radar and were easily avoided. Bergy bits and hard ice chunks from broken floes, invisible on the radar amongst all the clutter of brash ice surrounding the ship, if picked up by giant waves and hurled against the hull could pierce the steel as though it were cardboard. In addition to the watchers on the wings of the bridge there was now a man in the bows, peering ahead to search out the larger growlers and calling warnings back to the bridge through a loudspeaker system.

Down the sides of the ship moved a creamy-brown 'concrete mix' of sludge and ice pieces. Big water-rounded boulders of ice banged noisily against the hull and there was a continuous scraping sound as sharper pieces of broken floes tore at the steel. I had not been in anything like this since 1954 in *Kista Dan*.

The atmosphere on the bridge was dramatic. The captain was tense, barking out orders, asking questions and leaving the engine controls occasionally to peer anxiously at the radar screen. As the ship wallowed forward, with huge waves sweeping up astern, the wind, laden with snow, screamed through the rigging and coated the decks with ice. All the bridge windows were opaque with clotted snow; visibility was possible only through the round inserts of rapidly rotating glass that constituted the 'clear-view' screens.

The barometer trace, that had plunged steeply, was beginning to flatten out. It had fallen from 1018 bars to 966 in a little over 12 hours. By 0400 hours the wind had moderated to around 50 knots, so I returned to my cabin, lay down in my clothes and slept fitfully until 1100.

At 1230 the barometer was trying to rise but the wind was still 50-60 knots with gusts to 72. The waves were short and steep and the ocean was streaked with spume and spindrift. The captain was still busy on the bridge and again turned the ship to head back east into the storm. We rolled violently again but not as frighteningly as at the height of the storm.

I wrapped a still camera and a cine camera in transparent plastic bags and went out onto the deck to photograph the waves. On the foredeck two sailors struggled to tighten the retaining chains anchoring one of the DUKWs. The deck and rigging were sheathed in ice. I obtained a number of dramatic photos, then returned to my cabin to strip off my dripping windproofs and sponge off my face and beard.

At 1700 hours the swell lessened, probably because we were in the lee of the tip of the iceberg tongue, so the captain headed north to try to move away from the sea ice zone. Some very nasty bergy bits—as big as small houses—floated by and there were many almost-submerged 'growlers'. The chief engineer rigged a spotlight from the motor boat on the boat deck near the bridge, focusing it on the starboard side just forward of the bows. We were forced to turn east as the swell and the seas built up again.

After dinner the wind was still high and the seas very rough, but the huge swells of the hurricane had gone, the barometer was climbing steadily and those on the bridge were at last able to relax. I reflected that the only storms I could remember of this intensity were the 1954 hurricane (*Kista Dan*), the 100-knot storm faced by *Labuan* in 1951 and the storm from which *Wyatt Earp* sheltered behind the Balleny Island in 1948.

I reflected, also, that excessive caution can be dangerous. If the captain had proceeded at full speed immediately we had left Lewis Island, he would have passed out of the pack ice zone before the height of the storm and could have ridden it out in open water.

APPENDIX I

The Beaufort Scale of Wind Strengths

Beaufort number	Average wind velocity (knots)	Seaman's description of wind	Steel full-rigged ship, deep loaded
0	0	Calm	—
1	2	Light air	Steerage way only
2	5	Light breeze	2 knots in full sail
3	9	Gentle breeze	3-4 knots in full sail
4	14	Moderate breeze	5-6 knots in full sail
5	19	Fresh breeze	Can just carry royals and light staysails
6	24	Strong breeze	
7	30	Moderate (half) gale	
8	37	Fresh gale	
9	44	Strong gale	
10	52	Heavy (whole) gale	Main topsail only
11	60	Storm	Storm staysail or trysail only
12	80+	Hurricane	No canvas can stand

One knot = 1.85 km/h

APPENDIX II

Glossary of Ice Terms

Bergy bit:
A massive piece of sea ice or disrupted hummocked ice; also medium-sized piece of floating glacier ice. Generally less than 5 m above sea-level and not more than about 10 m across.

Beset:
Situation of a vessel surrounded by ice and unable to move.

Brash ice:
Small fragments of floating ice, not more than 2 m across; the wreckage of other forms of ice.

Cirque:
A roughly semicircular shallow but extensive recess scalloped out of the side of a maountain by previous glacier action.

Close pack ice:
Composed of floes mostly in contact, such that navigation is difficult even for specially constructed vessels and icebreaker assistance may be required.

Crack:
Any fracture or rift in sea ice not sufficiently wide to be described as a lead. It is usually possible to jump across a crack.

Fast ice:
Sea ice of greatly varying width which remains fast along the coast, where it is attached to the shore, to an ice wall, to an ice front, or over shoals, generally in the position where originally formed. Fast ice may extend 400 km from the coast.

Firn:
The imperfectly consolidated granular snow of the glaciers.

Floe:
A piece of sea ice other than fast ice, large or small. Light floes are anything up to a metre in thickness.

Frazil ice:
Fine spicules or plates of ice in suspension in water. The first stage of freezing, giving an oily or opaque appearance to the surface of the water.

Glacier:
A mass of snow and ice continuously moving from higher to lower ground or, if afloat, continuously spreading.

Glacier tongue:
An extension of a glacier, projecting seaward, and usually afloat.

Growler:
A piece of ice almost awash, smaller than a bergy bit.

Hummocked ice:
Sea ice piled haphazardly one piece over another, and which may be weathered.

Hummocking:
Process of pressure formation by which level ice becomes broken up into humps and ridges. When the floes rotate in the process it is termed 'screwing'.

Iceberg:
A large mass of floating or stranded ice broken away from a glacier or from an ice shelf. Often of considerable height (in any case more than 5 m above the level of the sea).

Ice blink:
Yellowish white luminous glare in the sky produced by the reflection on the clouds of pack ice or an ice sheet which may be beyond the range of vision.

Ice cover:
The amount of sea ice encountered; measured in eighths of the visible surface of the sea covered with ice.

Ice edge:
The boundary at any given time between the open sea and sea ice of any kind, whether drifting or fast. When wind or swell compact the ice edge it may be called a compacted ice edge; when dispersed, it may be called an open ice edge.

Ice rind:
A thin, elastic, shining crust of ice, formed by the freezing of sludge on a quiet sea surface. Thickness less than 5 cm.

Ice sheet:
A continuous mass of ice and snow of considerable thickness and large area.

Ice shelf:
A floating ice sheet of considerable thickness. Ice shelves are usually of great horizontal extent and have a level or gently undulating surface.

Katabatic wind:
Wind caused by the local gravitation of cold air down a steep slope.

Lead:
A navigable passage through pack ice; a lead may still be so called even if covered with young ice.

Névé:
Deposited snow that has consolidated and densified with the expulsion of most of the entrapped air bubbles.

Open pack ice:
Composed of floes seldom in contact and with many leads and pools.

Pack ice:
Any area of sea ice other than fast ice, no matter what form it takes or how it is disposed.

Pancake ice:
Pieces of newly formed sea ice usually approximately circular in shape with raised rims, and from one-third to 2 m in diameter.

Polynya:
A large area of open water, surrounded by sea ice and found in the same region every year. One side is sometimes formed by the coast. Polynyas commonly occur off the mouths of big rivers.

Pressure ice:
A general term for sea ice which has been squeezed, and in places forced upwards when it can be described as rafted ice, hummocked ice or pressure ridge.

Pressure ridge:
A ridge or wall of hummocked ice where one floe has been pressed against another.

Rafting:
The overriding of one floe on another, the mildest form of pressure. Hence 'rafted ice'.

Rotten ice:
Sea ice which has become honeycombed in the course of melting, and which is in an advanced state of disintegration.

Serac:
An irregular-shaped pinnacle of ice on a glacier, formed by the intersection of crevasses.

Sludge:
A later stage of freezing than frazil ice, when the spicules and plates of ice coagulate to form a thick soupy layer on the surface. Sludge reflects little light, giving the sea a matt appearance.

Tabular berg:
A flat-topped iceberg. Most tabular bergs form by breaking from an ice shelf and show horizontal firm snow layers.

Tellurometer:
Trade name of a device used to accurately measure distances between objects of up to about 30 kilometres. It used technology similar to that of radar.

Tide crack:
The fissure at the line of junction between an immovable icefoot or ice wall and fast ice, the latter being subject to the rise and fall of the tide.

Tongue:
A projection of the ice edge up to several kilometres in length, caused by wind and current.

Winter ice:
More or less unbroken, level sea ice of not more than one winter's growth originating from young ice. Thickness from 15 cm to 2 m .

Young ice:
Newly formed level ice in the transition stage of development from ice rind or pancake ice to winter ice. Thickness from 5 to 15 cm.

Source

Terence Armstrong and Brian Roberts, *Illustrated Ice Glossary*, Scott Polar Research Institute, Cambridge, 1956.

INDEX